A Course in Organic Chemistry

edited by SIR ROBERT ROBINSON, O.M., F.R.S.

ADVANCED SECTION Volume XXVII

INDOLE ALKALOIDS

INDOLE ALKALOIDS

An Introduction to the Enamine Chemistry of Natural Products

by

W. I. TAYLOR

CIBA Pharmaceutical Company

PERGAMON PRESS

OXFORD · LONDON · EDINBURGH · NEW YORK
TORONTO · PARIS · BRAUNSCHWEIG

Pergamon Press Ltd., Headington Hill Hall, Oxford
4 & 5 Fitzroy Square, London W.1

Pergamon Press (Scotland) Ltd., 2 & 3 Teviot Place, Edinburgh 1

Pergamon Press Inc., 44–01 21st Street, Long Island City, New York 11101

Pergamon of Canada, Ltd., 6 Adelaide Street East, Toronto, Ontario

Pergamon Press S.A.R.L., 24 rue des Écoles, Paris 5e

Friedr. Vieweg & Sohn, Verlag, Postfach 185, 33 Braunschweig,
West Germany

First edition 1966

Library of Congress Catalog Card No. 65–26890

Printed in Great Britain by
Spottiswoode, Ballantyne & Co. Ltd., London and Colchester

(2487/66)

TO MY
MOTHER AND FATHER

NOTE BY EDITOR

THE volumes in the Advanced Section of this Course in Organic Chemistry have an independent value as short monographs by experts in the respective fields treated. They are therefore offered in advance of the General Sections of the Course, even though this implies a greater prior knowledge on the part of the student.

The present volume breaks new ground from the point of view of presentation, since the special type of reactivity of the indole system is placed in the forefront and related to the properties of the alkaloids and their biogenesis. This treatment is original and it will certainly be found instructive.

CONTENTS

CHARTS

TABLE

PREFACE

THIS book owes its genesis to a request, made by Sir Robert Robinson on the occasion of his seventy-seventh birthday, to write a monograph on indole alkaloids. I have stated concisely my views on the chemistry and interrelationships between indole alkaloids which should be of value to teachers, a challenge to students and even interesting to practitioners of the art.

I owe much to the management of CIBA Pharmaceutical Company for creating the environment which made possible certain contributions to the chemistry discussed on the subsequent pages.

I am indebted to Dr. E. Schlittler who read the entire manuscript and made many helpful suggestions. My thanks are also due to Miss Susanne Hay who typed the manuscript and to Charles Gemenden for his help in the proofreading.

W. I. T.

Summit, N.J.
October, 1964

INTRODUCTION

AT THE dawn of experimental organic chemistry just over a century and a half ago, Sertürner isolated morphine from opium. Although he was not the first man to do so he did recognize it to have basic properties and referred to it as a "vegetable alkali". Meisner in 1818 proposed that such vegetable alkalis should be called "alkaloids". This noun became accepted and today the word generally denotes a basic, physiologically active nitrogen heterocycle of some complexity obtained from plant sources. This definition however does not prevent the chemist from considering in the same context and for reasons of chemical or physiological properties simple primary amines such as the cactus hallucinogen, mescaline, the neutral mitotic poison, colchicine from the autumn crocus; the nitrophenanthridine, aristolochic acid from *Aristolochia sipho*; and even complex pharmacologically active bases of animal origin, *inter alia*, dehydrobufotenine from the toad and samandarine from the salamander.

In this book we shall concern ourselves with the chemistry of selected alkaloids which contain indolic or closely related nuclei and greater emphasis will be placed on developing the chemistry of the complex indoles than rigid proofs of their structures or syntheses. Some five hundred of these bases have been obtained from about three hundred plants mostly of the family *Apocynaceae*. The alkaloids fall into two broad classes in which a tryptophan residue or its equivalent is either (a) in combination with a ten carbon moiety and this forms the greater majority of the bases or (b) modified slightly by alkylation, ring closure or fusion to an anthranilic or mevalonic acid residue.

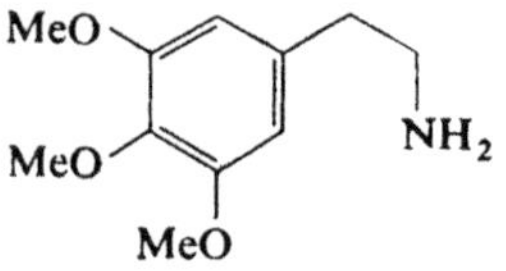

mescaline

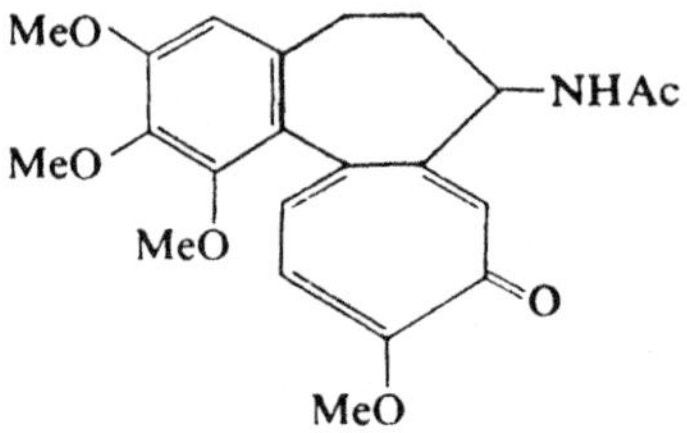

colchicine

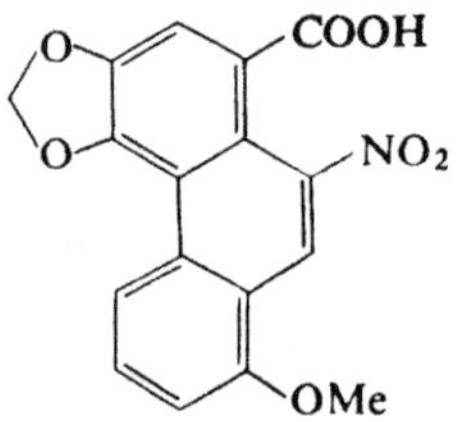

aristolochic acid

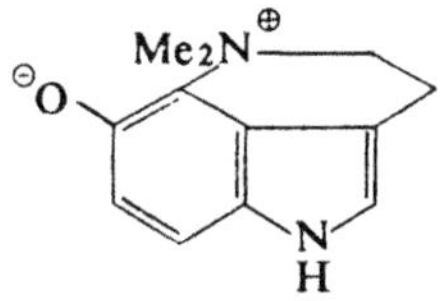

dehydrobufotenine

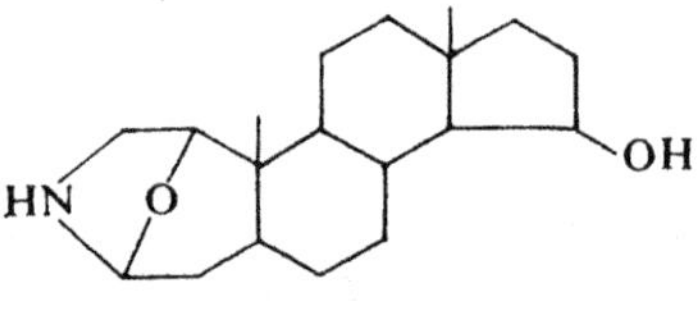

samandarine

Elucidation of the structures of the first group of alkaloids was slow until the advent of commercial recording ultraviolet and infrared spectrometers allowed the first major break through. We are presently witnessing a second revolution in the conduct of the art brought about by the availability of protons mappers (nuclear magnetic resonance machines) and mass spectrometers. The application, however, of the greatest ultimate promise is structure determination by means of the interpretation of X-ray diffraction data.

Indole alkaloids although amongst the earliest known natural products did not readily lend themselves to structural elucidation and as a result there was an accumulation of experimental facts which outpaced understanding. Like other natural products of unknown structural type, an indole alkaloid was given a trivial name usually based on that of the first plant from which it was isolated. In order to be able to describe what he had done, the chemist used the name of his alkaloid as a root word and added prefixes to indicate in a shorthand way what he had done or thought he had done. Prefixes such as *pseudo*, *iso* and *neo* (like, similar to, isomeric), *apo* (something detached), *anhydro* (minus water), *nor* (without, usually minus a methylene), *seco* (a ring opening) and *chano* (a possible ring opening but may not be so), came into being and were in some cases not wisely used. A good example of the value of this system lies in the naming of the product obtained when a Hofmann elimination was attempted on strychnine methosalt in methanolic potassium hydroxide. The product was not the methine base but rather a rearrangement product of strychnine which had retained the N-methyl and undergone a displacement reaction with methoxide ion and, therefore, a ring opening (equivalent to adding two hydrogens) had occurred. The name chosen for the isolate was N-methylmethoxydihydroneostrychnine. Today because of the rapidity with which structures can be determined a more systematic nomenclature is gradually being evolved.

In this book a mastery of neither the structural formulae nor the trivial names is required of the reader since the chemistry can be intelligently followed without their use. However, cultivation of a memory for them will make the transition to an encyclopedic text easier. The numbering system is coupled to the biogenetic theme and attempts (with the exception of the simple derivatives of tryptophan) to give equivalent atoms the same number irrespective of the particular heterocycle in which they find themselves.

The reader is assumed to have a good grounding in organic reaction mechanisms and therefore can rationalize many of the details for himself. If he should not be too familiar with enamine chemistry his attention is directed towards Chart 3.5 which is the key to the understanding of a large part of the chemistry he is about to read. This Chart is also reproduced as a foldout on the back page.

Although the book is intended as a serious introduction to a study of indole alkaloids it could be complementary to any graduate course in organic chemistry. It charts the classical and current approach to the examination of natural products and the summaries of possible biogenetic relationships are predictions of chemistry which will be realized in the laboratory within the next few years irrespective of their biosynthetic validity. The Charts can be used independently of the text which is why they have been listed in the Table of Contents.

GENERAL REFERENCES

HENRY, T. A., *The Plant Alkaloids*, J. & A. Churchill, London, 1949.

MANSKE, R. H. F., *The Alkaloids*, Vols. 1–8, Academic Press, New York, 1950–1965.

BOIT, H. G., *Ergebnisse der Alkaloidchemie bis 1960* unter besonderer Berücksichtigung der Forschritte seit 1950, Akademie-Verlag, Berlin, 1961.

Chemical Society, London, *Annual Reports*.

WILLAMAN, J. J. and SCHUBERT, B. G., *Alkaloid Bearing Plants and Their Contained Alkaloids*, U.S. Dept. Agr., Techn. Bull. **1234**, 1961.

CHAPTER I

ISOLATION AND CHARACTERIZATION OF INDOLE ALKALOIDS

THE following remarks, although written specifically with indole alkaloids in mind, are good for alkaloids in general and basically true for any isolation problem. At some time during the workup of a plant extract advantage is taken of that property of the looked for substance(s) which will enable its separation from the other products. For an alkaloid one normally takes advantage of its basicity unless some other incidental property can be effectively used. In its simplest form an extract of the plant in a water immiscible solvent is washed with aqueous acid which upon subsequent basification affords the alkaloids.

This simple and direct method has possible liabilities since the strong acid which may have to be used to get out the weak bases may induce undesirable changes in the alkaloid mixture (e.g. eburnamine, p. 123). Also extraction of the plant with the water immiscible solvent probably would have left behind the quaternary bases, N-oxides, amino acids and perhaps even phenolic bases.

A useful and systematic approach is to make an extraction with petroleum ether to remove fats and waxes keeping in mind that some bases can be extracted with such a solvent and to follow this up with wet benzene. The benzene solution should contain the tertiary bases and washing it with buffered solutions of decreasing pH will yield crude alkaloids divided roughly into groups of similar basic strengths. The aqueous phase resulting from the wet benzene extraction would be expected to contain any alkaloid salts, the N-oxides, quaternary bases, amino acids and perhaps phenolic bases.

A simple version of this scheme of commercial importance is illustrated by obtention of reserpine (pK_a 6·06) from the roots of *Rauwolfia vomitoria.* The ground roots are extracted with benzene which is washed with aqueous acetic acid to remove the strong bases, and concentrated whereupon the reserpine crystallizes out.

Occasionally isolation of a particular alkaloid is facilitated by forming a water insoluble salt, nitric, perchloric, sulphuric and oxalic acids have often been used for this purpose but this does not form part of any systematic procedure except for the removal of water soluble bases from aqueous solution. Thus for quaternary salts picric acid and reinecke salts have been widely used. N-oxides also form water insoluble derivatives with many acids but this problem can usually be dealt with by reduction back to the parent tertiary base which is then soluble in organic solvents.

In order to follow the course of an alkaloid isolation the classical chemist developed many precipitation and chromogenic methods. They either depended on the water insolubility and colour of complex acid salts or the generation of characteristic colours with oxidizing agents under defined conditions. For indole alkaloids some colour reactions are specific to a particular indolic system and in a very crude mixture such a test is often more useful than attempting to measure and interpret an ultraviolet absorption spectrum. Useful colours are generated by ceric sulphate, or in sulphuric acid containing a trace of an oxidizing agent, e.g. ferric chloride (Keller), sodium nitrite (Arnold-Vitali), sodium molybdate (Fröhde) and potassium dichromate (Otto).

For precipitations the most well known reagents are Mayer's (potassium mercury iodide), Sonnenschein's (phospho-molybdic acid) and occasionally picric acid. Today these reagents have a new use as detectors in paper and thin layer chromatography.

Once the crude alkaloids have been obtained separation methods are applied which take advantage of previous experience and modern methodology such as column, partition or ion exchange chromatography, countercurrent distribution and the like. The results of the fractionation can be followed by all or as many of the physical methods as need be brought to bear to the problem.

The isolated pure base can be characterized via its melting point, optical rotation (or optical rotatory dispersion curve), ultimate analyses and by the formation of derivatives. All of these measurements were very important up to about 1950 since these were the only means available for comparing compounds. Today many of these pointer readings have diminished in value because more accurate comparisons can be made by a multiplicity of physical methods which sometimes result in a complete proof of structure. The ultraviolet absorption spectrum defines the chromophoric moiety and the infrared spectrum besides "fingerprinting the molecule" by its wealth of bands, detects various functionalities in particular, carbonyl groups and hydrogen attached to nitrogen or oxygen. Nuclear magnetic resonance spectroscopy yields information from which the proton topography can be deduced. Then there is the mass spectrum the interpretation of whose line rich spectrum often allows the entire structure of the alkaloid to be deduced.

With the advent of computers it has become possible to reduce the interpretation of the X-ray diffraction data of a single crystal containing a heavy atom (e.g. the alkaloid iodide, bromide, brosylate) to a routine. This method is absolute and requires no assumptions as to the structure and it will eventually be used by the chemist as naturally as he has accepted all the other aids which have come his way. It has already enriched our knowledge by solving problems for which there was either no chemical method available or simply not enough material for a chemical examination.

With the structure of the indole alkaloid largely defined by use of the above methods the chemist can operate more meaningfully with small amounts of material to complete the structure proof. This can take the form of either partial degradation to or partial synthesis from a known base, which at the same time settle the absolute stereochemistry.

With the structure, stereochemistry and conformation now relatively easy to determine and the degradative process eliminated there remains the challenge of synthesis and it must be by a rapid, high yielding stereospecific (or selective) route. There are as yet few landmarks in this area but it is one to which an increasing amount of attention is being paid.

GENERAL REFERENCES

BUDZIKIEWICZ, H., DJERASSI, C., and WILLIAMS, D. H., *Structure Elucidation of Natural Products by Mass Spectrometry*, Vol. I, Alkaloids, Holden-Day, San Francisco, 1964.

DJERASSI, C., *Optical Rotatory Dispersion: Applications to Organic Chemistry*, McGraw-Hill, New York, 1960.

HAMERSLAG, F. E., *The Technology and Chemistry of Alkaloids*, Nostrand, New York, 1950.

HESSE, M., *Indolalkaloide in Tabellen*, Springer-Verlag, Heidelberg, 1964.

JACKMAN, L. M., *Applications of Nuclear Magnetic Resonance Spectroscopy in Organic Chemistry*, Pergamon Press, London, 1959.

MACEK, K., HAIS, I. M., GASPARIC, J., KOPECKY, J., and RABEK, V., *Bibliography of Paper Chromatography 1957–1960 and Survey of Applications*, Publishing House of Czechoslovak Academy of Sciences, Prague, 1962.

MATTHIEU, J. P. and JANOT, M.-M., *Selected Constants IV, Optical Rotatory Power and Alkaloids*, Pergamon Press, London, 1961.

NAKANISHI, K., *Infrared Absorption Spectroscopy—Practical*, Holden-Day, San Francisco, 1962.

NEUSS, N., *Physical Data of Indole and Dihydroindole Alkaloids*, Eli Lilly and Company, Indianapolis, Indiana, 1961.

NYBURG, S., *X-ray Analysis of Organic Structures*, Academic Press, New York, 1961.

SCOTT, A. I., *Interpretation of the Ultraviolet Spectra of Natural Products*, Pergamon Press, London, 1963.

SIXMA, F. L. J. and WYNBERG, H., *A Manual of Physical Methods in Organic Chemistry*, John Wiley and Sons, New York, 1964.

CHAPTER II

THE ORIGIN OF INDOLE ALKALOIDS

THE distribution of the complex indole alkaloids in plants will not be discussed except to point out that they occur most frequently in the family *Apocynaceae* and to a lesser extent in the *Asclepiadaceae*, *Loganiaceae* and *Rubiaceae* and it is worth remembering that these families stand close together in the phylogenetic charts of the taxonomists. The simpler indole derivatives are of much wider distribution but likewise confined largely to the Dicotyledons.

On the chemical side it has long been the practice of the chemist to attempt to discern common denominators within a class of natural products by a process which can best be regarded as a visual dissection of the structural formulae. The most well-known example of such an exercise has been the recognition of repeating branched chain five carbon units in terpenes and the subsequent use of this observation in reverse in order to frame structures which could then be tested against the experimental results. Mevalonic acid has turned out to be the biochemical equivalent of the hypothetical branched five carbon unit. In the case of indole alkaloids not quite so much progress has been made.

There can be recognized in the majority of indole alkaloids a 3-β-aminoethylindole moiety and in agreement with an early hypothesis it has been shown that this can be derived from tryptophan. It is the source of the remaining portion, a nine or ten carbon fragment, which is presently obscure in spite of a number of sophisticated suggestions and a lot of experimental work. Whatever the progenitor of the ten carbon residue is, it can

CHART 2.1

PRECURSORS OF THE COMPLEX INDOLE ALKALOIDS

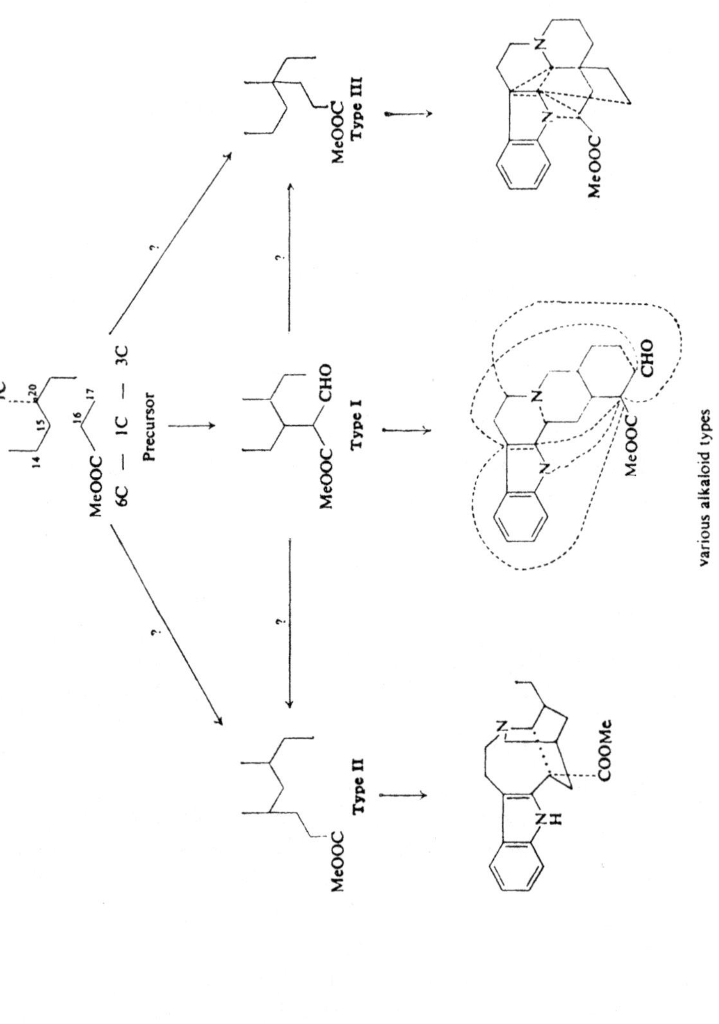

be seen to be divisible into a linear six carbon plus one carbon residue (6C–1C) linked in one of three ways to a three carbon unit (or 2C if the carboxyl is lost) as shown in Chart 2.1, where some idea of diversity of indole alkaloids is given. Parenthetically it should be noted that Type II and III alkaloid skeleta are theoretically derivable from Type I via four membered ring intermediates, i.e. fission of the C-15, C-16 bond and reclosing of C-17 with either C-14 or C-20, but such rearrangements are presently unknown. If Type I precursor is in actual fact being converted into Types II and III it is not known whether it happens before or after combination with tryptophan. Some alkaloids are known which lack either the 3C-unit or the C-1 unit.

The classical view of the origin of the indole alkaloids was predicated almost entirely upon the structure of a single alkaloid yohimbine (even before the site of its carboxyl was known). This alkaloid was considered to stem from tryptophan, phenylalanine or its equivalent and formaldehyde via two Mannich condensations (α-condensation, Chart 2.2). The successful preparation of a model under so-called "physiological conditions" supported this idea but how either the carboxyl group was to be introduced or how reduction of the aromatic ring E was to be accomplished was never seriously discussed. In order to account for the origin of strychnine the ingenious suggestion was made that instead of an α-condensation process there was one at the β-position with subsequent fission of the 3,4-dihydroxyphenyl moiety as illustrated in Chart 2.2. This concept was applied almost immediately to other indole alkaloids as well as to the bisisoquinolinyl alkaloid, emetine, which has in its core the Type I precursor (Chart 7.1). Despite the manifold variety of indoles now known it is only the yohimbines and their dehydrogenated equivalents which have a carbocyclic ring E. It is possible that these compounds are formed by ring closure of a Type I precursor and that the earlier view was the result of a historical accident, viz. that the structure of yohimbine became known just simply because it was most amenable to the then developed degradative methods. This hypothesis may not be valid since the administration of ^{14}C-labelled phenylalanine or

CHART 2.2
TWO MODES OF CONDENSATION FOR TYPE I ALKALOIDS

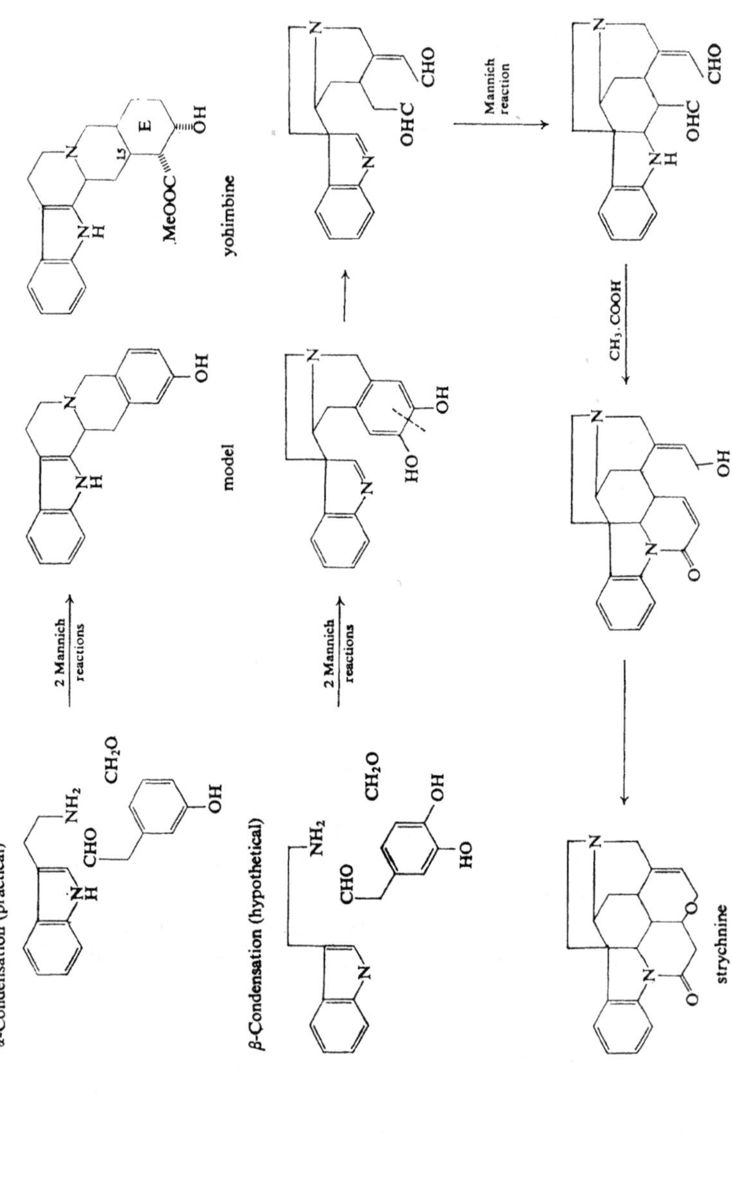

tyrosine into *Rauwolfia serpentina* plants has resulted in only a low and nonspecific incorporation into alkaloids such as serpentine and ajmaline.

An alternative to the aromatic amino acids hypothesis is one which invokes their proximate precursor, viz. a hydrated prephenic acid in which the potential ring E of yohimbine is already in a reduced state and has in addition the carboxyl. An important conclusion arising out of this theory was the prediction of a uniform stereochemistry in indole alkaloids of Type I at the carbon equivalent C-15 of yohimbine. This hypothetical biogenesis—the prephenic acid theory is illustrated in Chart 2.3.

A later version of this theory starts with a ring opened unit plus a one carbon fragment so that the genesis of the Type II and III precursors could be rationalized, this was considered to take place after combination with the indolic moiety as sketched in Chart 2.3. The prephenate hypothesis has received only one test and that was to feed 2-^{14}C-alanine to *Rauwolfia serpentina* plants. The amino acid was expected to be converted in part to pyruvic acid thence to prephenic acid, etc. The actual uptake into ajmaline was exceedingly poor and randomized since only 2% of the incorporated radioactivity was at the predicted position, therefore at the present time support for this theory is lacking. It should be noted that in Chart 2.3 it is indicated that the basic nitrogen of these alkaloids is not necessarily derived from that of tryptophan and it has been emphasized that the secoprephenate formaldehyde grouping can be written in a number of other equivalent ways.

In the present state of biogenetic theory, it cannot be assumed that the combination of the ten carbon precursor and tryptophan must always lead to a basic compound. The recent isolation of certain aspidosperma alkaloids which owe their basicity and therefore the reason they were isolated to the indoline nitrogen may mean *inter alia* that the analogous yohimbine derivative (lactam carbonyl group at C-5 in Chart 2.3) is also a natural product.

Quite a few other natural compounds are now being recognized in which the Type I precursor can be discerned. Among others,

CHART 2.3
PREPHENIC ACID THEORY OF INDOLE ALKALOID BIOSYNTHESIS

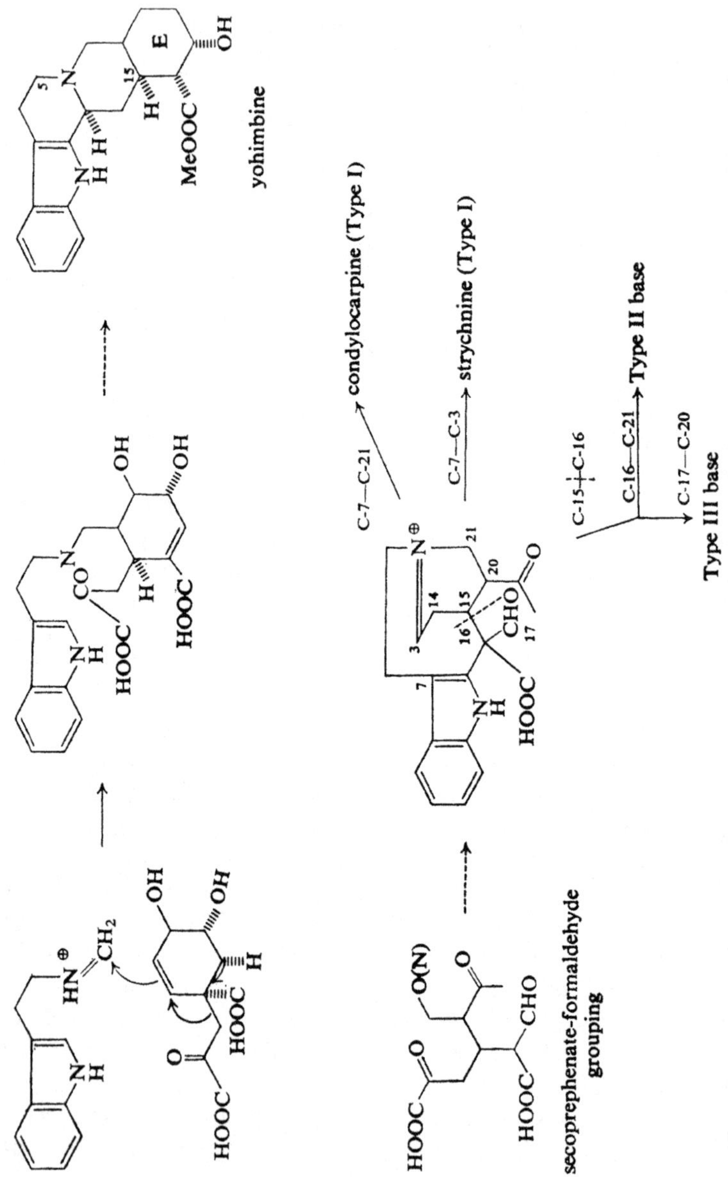

CHART 2.4

CYCLOPENTANO TERPENIC GLYCOSIDES AND ALKALOIDS

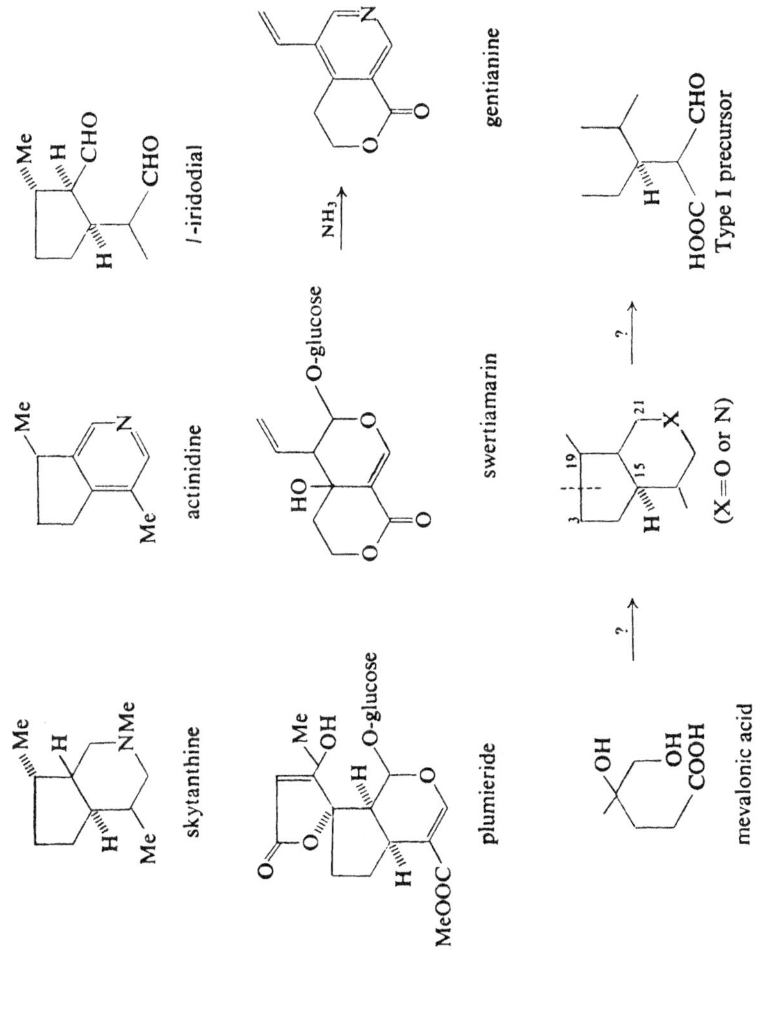

there is gentianine [perhaps natural, but in the case of *Anthocleista procera* it is an artefact isolated only if ammonia is used in the workup] and the cyclopentanoterpenic glycosides and alkaloids (Chart 2.4). It has been shown, at least for plumieride that the system is derived by the head to tail linkage of two mevalonic acid residues. If a split is imagined to occur between carbons 3 and 19 (Chart 2.4) in the cyclopentano terpenic moiety then the carbon pattern of Type I precursor is generated. A single practical test using mevalonic 2-^{14}C-acid in *Rauwolfia serpentina* has afforded inactive ajmaline.

One of the possibilities for the theory has been tested using the assumption that it is equivalent to 3 acetic, 1 formic and 1 malonic acids and experimentally the predicted specific labelling has been claimed for ajmaline and serpentine.

Although the question of the germinal precursors, apart from tryptophan of the indole alkaloids has not been settled it will be seen throughout this book that the chemist has learned to mimic and to reverse many of the plausible terminal steps. This combination of real knowledge and biogenetic theory has resulted in some synthetic, degradative and interconversion sequences of great elegancy.

SUGGESTED READING

ROBINSON, Sir ROBERT, *Structural Relations in Natural Products*, Clarendon Press, Oxford, 1955.

SCHLITTLER, E. and TAYLOR, W. I. A view of progress in the chemistry of indole alkaloids, *Experientia*, Basle, **16**, 244 (1960)

BATTERSBY, A. R. Alkaloid biosynthesis, *Quarterly Reviews*, **15**, 259 (1961).

TAMELEN, E. E. VAN. Biogenetic type syntheses of natural products, *Progress in Chemistry of Natural Products*, **19**, 242 (1961).

BERNFELD, P. (Ed.) *Biogenesis of Natural Products*, Pergamon Press, 1964.

WENKERT, E., Biosynthesis of indole alkaloids. The aspidosperma and iboga bases, *J. Amer. Chem. Soc.*, **84**, 98 (1962).

CHAPTER III

BASIC CHEMICAL CONSIDERATIONS

NOT having the advantages of current methodology the classical chemist approached the examination of an indole alkaloid in the following ways:

1. He tried to establish the nature of the molecular framework by converting it into identifiable fragments usually aromatic compounds. This was often accomplished by vigorous oxidation, dehydrogenation or pyrolysis under basic conditions.
2. He detected and estimated the functional groups.
3. He searched for interrelationships between the functional groups and their sites on the molecule.
4. He worked, if possible, upon related alkaloids, since the solution of the structure of one might greatly simplify work on the others.
5. He attempted to use one or more of the functionalities as the starting point for a stepwise degradation.

Progress was slow because many of the indole alkaloids were not susceptible to the then known methods and they were not always available in sufficient quantity to find the correct experimental approach. In fact progress was often disappointingly slow even when the alkaloid was commercially available, e.g. strychnine discovered in 1818 was the subject of intensive chemical work over sixty years before its structure was definitively settled in 1947. During much of that period the chemist had to try to understand much of his chemistry through his interpretation of the differences

Chart 3.1
Strychnine and yohimbine and their fusion products

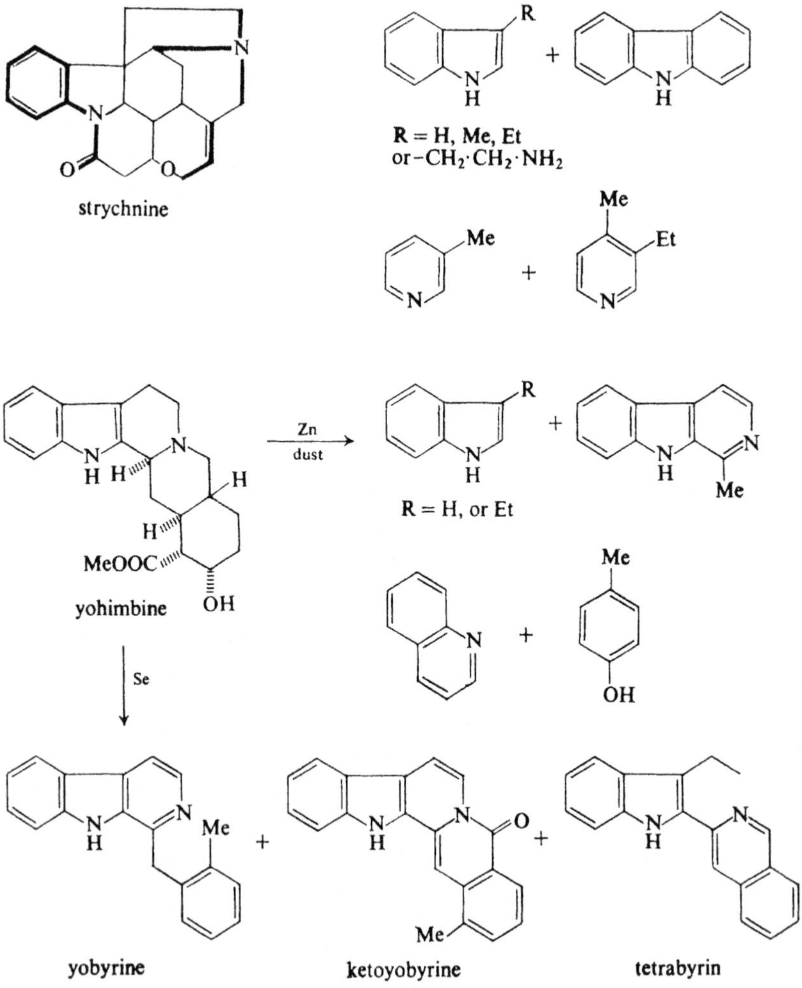

between elemental compositions of compound and derivative as revealed by the combustion figures. The introduction of the Pregl microcombustion method was an important instrumental advance in its day because it released so much more material for bench work. The experimentalist groped for understanding by intuition, often attaining brilliant levels, and above all a constant search for comparable results in published literature, but he was not always able to find an analogy because he was a creator of new chemistry. Another difficulty stemmed from his not really considering his molecule as a three dimensional object with the resultant structural and mechanistic implications. In fact, even the study of absolute stereochemistry was not usually pursued by those who specialized in degradative work.

On no compound can the development of this phase of organic chemistry be better understood than by reading the chemistry of strychnine where initially experiments were carried out empirically to be followed by more rational approaches as knowledge was accumulated. The final sophisticated experiments were conceived to confirm or deny good working hypotheses. This later phase was greatly simplified by the advent of the early ultraviolet and infrared spectrophotometers and in the end the structure was completely and independently confirmed along with absolute stereochemistry by the X-ray method. This structure for strychnine was proved again in the middle fifties by a total stereospecific synthesis and a chemical solution for absolute stereochemistry was provided for the first time in 1962.

Today the molecular formula and the obvious chromophoric moieties along with the type of double bond can be established rapidly by techniques which allow by and large for complete recovery of the alkaloid in question, e.g. for strychnine (Chart 3.1) the units shown by the more heavily shaded positions. This sort of information was available in the past via pyrolytic methods and two examples are given for strychnine and yohimbine in Chart 3.1. It can be seen that despite the so-called brutality of the method it did provide accurate structural information the problem being, of course, that in hindsight it is more obvious.

The modern equivalent of this technique is electron impact fragmentation (mass spectroscopy).

With strychnine the basic nitrogen although undergoing Hofmann, Emde and von Braun reactions did not furnish compounds suitable for further degradation but these studies generated

CHART 3.2
AKUAMMICINE HEATED AT 140° IN METHANOL

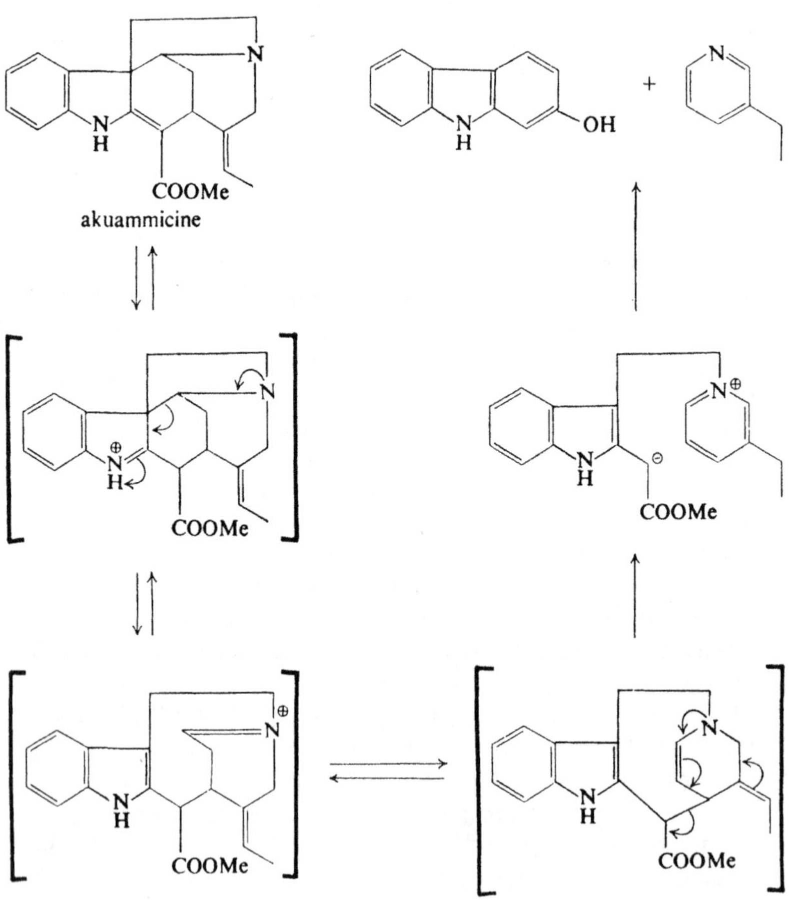

chemical results which had to be compatible with any serious structural proposal. It was the fission of the double bond and subsequent steps which enable a relationship to be established between the acylaniline residue and the basic nitrogen. If more effort had been put into continuing this systematic approach the structure of strychnine might have been established rather sooner than it was. We shall learn in later chapters that in general, degradative routes for dihydroindole alkaloids have not been as easy to find as ones for indole bases. There are exceptions to all rules and a particularly beautiful example lies in the fission of akuammicine in methanol at 140° to afford 2-hydroxycarbazole and a 3-ethylpyridine but it is not possible to deduce the structure of the parent alkaloid from these parts alone but their origin can be easily explained on the basis of the established formula.

Selenium dehydrogenation of indoles has on the other hand been a tool of great usefulness giving in many cases aromatic products containing all the cyclic carbons and the nitrogens of the original, see for example, yohimbine in Chart 3.1, leaving the site of the remaining functional groups, stereochemistry and synthesis to be considered. Another good example is the case of ibogaine the products of whose potassium hydroxide fusion and selenium dehydrogenation allow the deduction of the correct formula, except for the configuration of the C-ethyl (Chart 3.3).

In the early fifties by using the recording spectrophotometers the chemist was able to approach his problem in a new way, viz. by starting off with a known chromophoric system and modifying it to yield other known chromophoric systems. In this way he was able to map increasingly larger portions of the molecule under study. If the reactions were reversible he stood to gain in addition valuable stereochemical information. Some of the basic operations for tetrahydro-β-carboline alkaloids are indicated in Chart 3.4 and their success or failure in a practical test could be a combination of stereochemistry and/or the absence of the depicted system.

The oxidation with mercuric acetate is diagnostic for the stereochemistry at C-3, viz. it will proceed if the proton is *trans* to the electron pair of N-4, i.e. a *trans*-fused indoloquinolizidine.

Chart 3.3
THE POTASSIUM HYDROXIDE AND SELENIUM DEHYDROGENATION PRODUCTS OF IBOGAINE

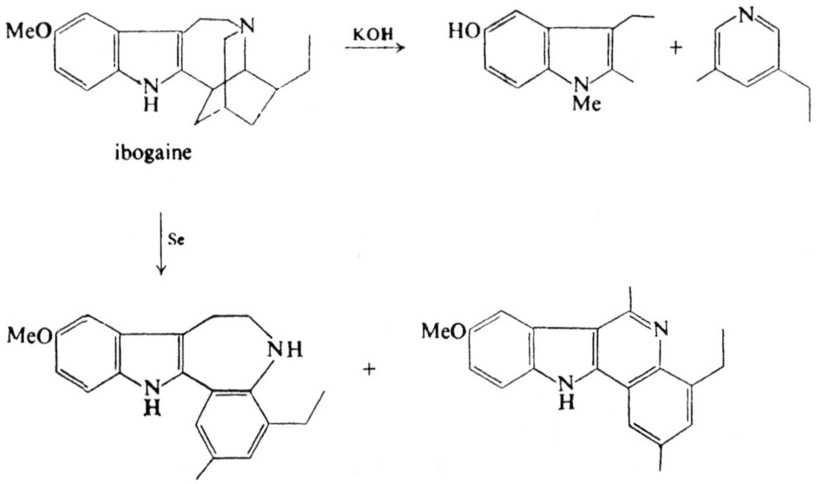

This stereochemical question and its control was much studied in the case of the complex tetrahydro-β-carboline alkaloids (Chapter V). Today from a purely structural point of view the mass spectrometric method can establish the presence of this ring system (rings ABC in Chart 3.4) on a few hundredths of a milligram.

A second aspect of the chemistry of the indolic nucleus lies in taking advantage of its initial nucleophilic reactivity to realize transformations of synthetic and degradative usefulness. This reactivity towards electrophiles is sometimes a problem, e.g. the sensitivity of many Type II alkaloids towards autoöxidation. The valuable transformations are sketched in Chart 3.5 for both condensations at the indolic α and β positions. α-Methyleneindoles turn out to be important intermediates in the formation and degradation of dimeric alkaloids, e.g., voacamine (p. 102) and vincaleukoblastine (p. 112). Although it is known that protonation, at least in simple indoles occurs predominantly on the β position we do not know in many cases whether the primary condensation

Chart 3.4

OXIDATIVE REACTIONS OF TETRAHYDRO-β-CARBOLINES AND RESULTS OF ELECTRON IMPACT

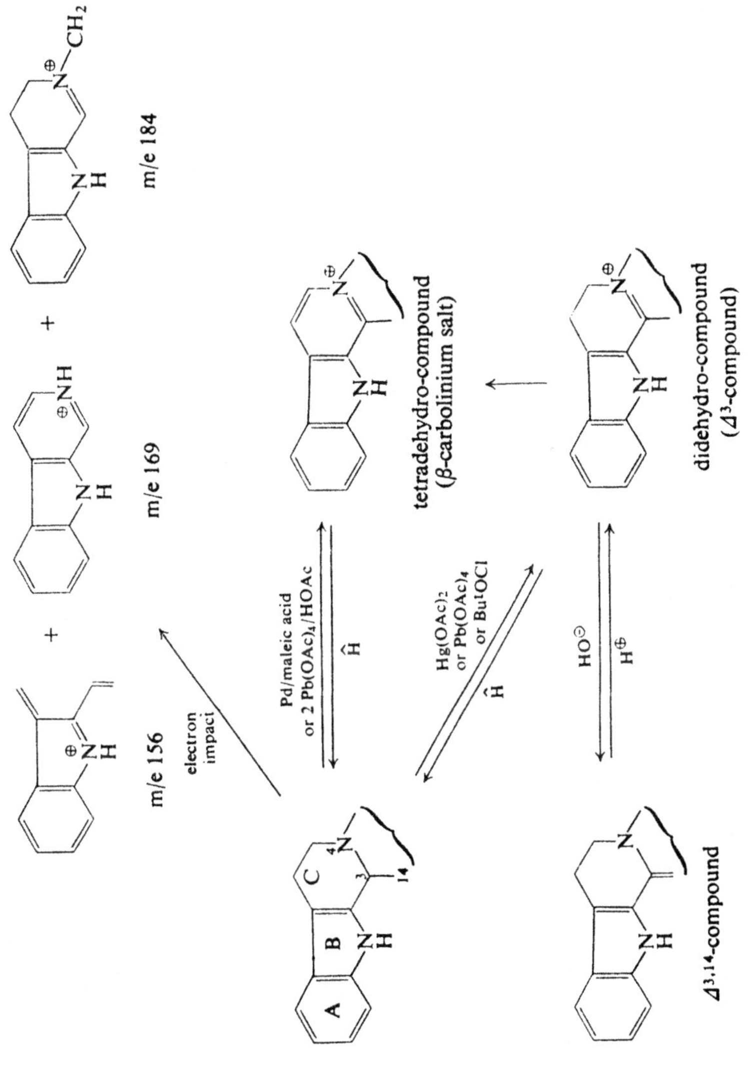

product of say a Pictet–Spengler or Bischler–Napieralski condensation may not rearrange from the α to the β position or *vice versa* before yielding the end product. In any event the final products of any of the reactions which are sketched are a combination of the addenda and the experimental conditions. Examples of some of these reactions are listed below and others are given in later chapters.

CHART 3.5
IMPORTANT REACTIONS OF INDOLES (SCHEMATIC)

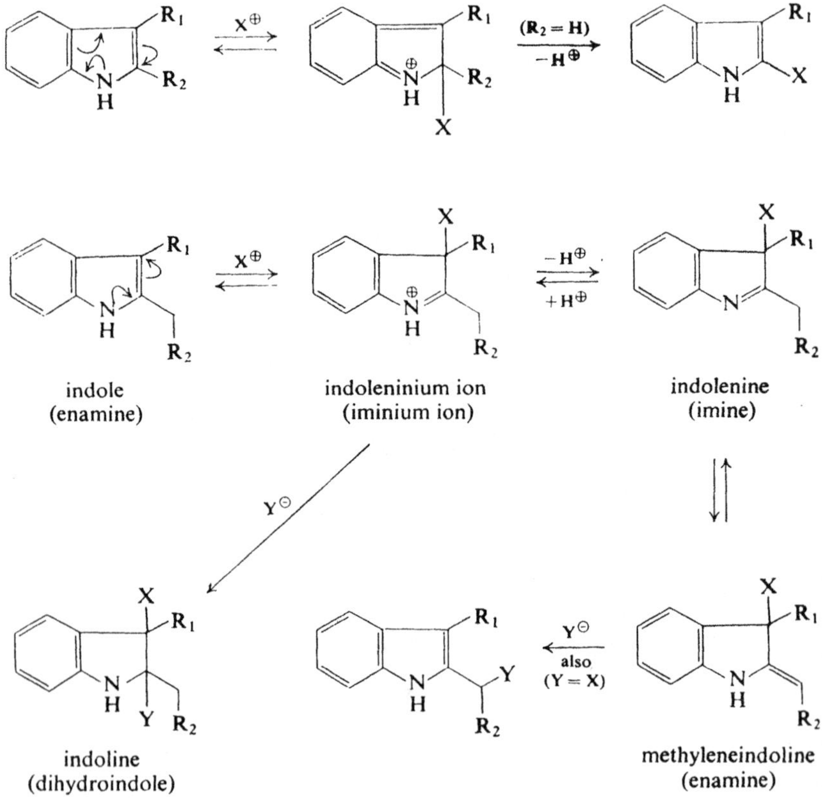

1. Mannich Condensation (Pictet–Spengler Condensation)

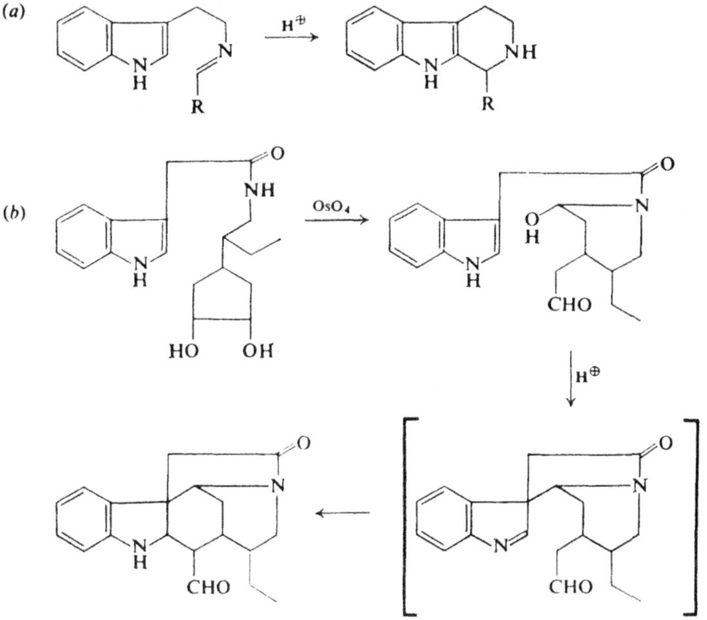

2. Bischler–Napieralski Condensation

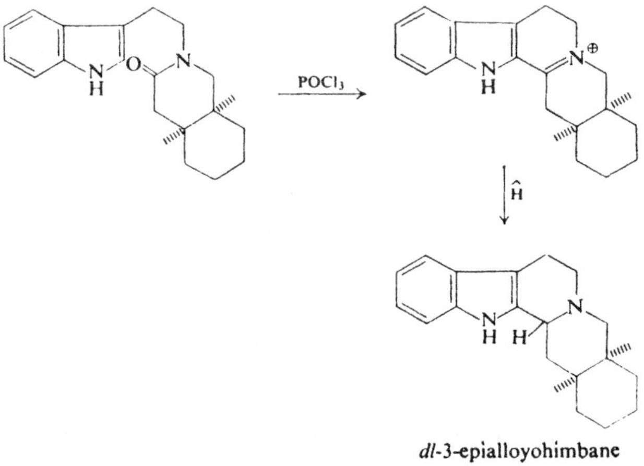

dl-3-epialloyohimbane

3. Acid catalysed equilibration of certain yohimbane derivatives at C-3 is mediated by one or more of the following equilibria the driving force being toward the direction of greater thermodynamic stability.

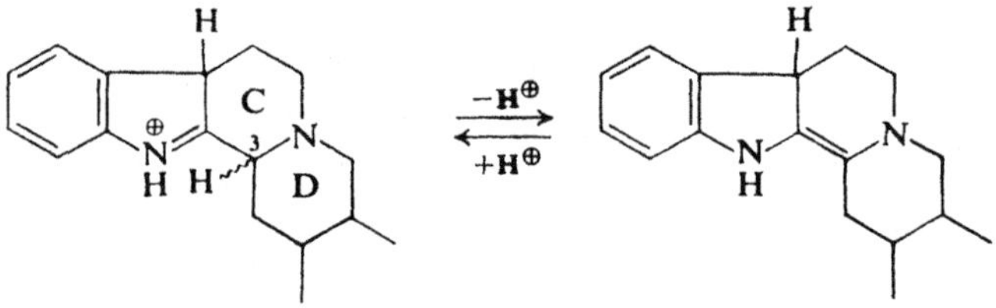

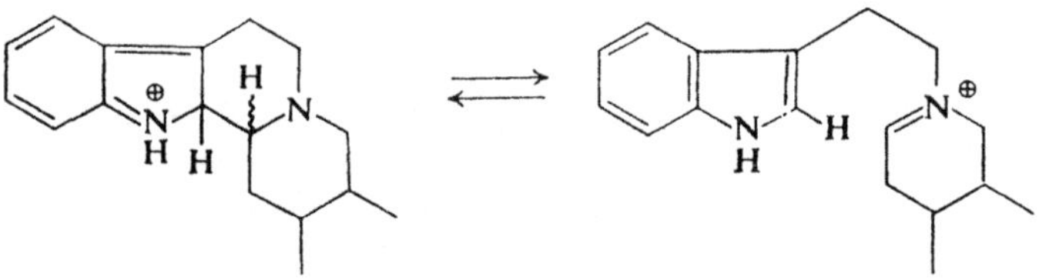

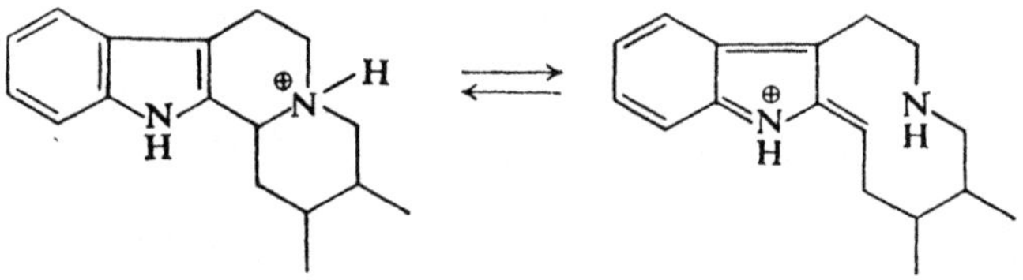

4. Autoöxidation products and their transformation products are well known for Type II and III alkaloids. For Type I alkaloids autoöxidation usually produces dehydro- and tetradehydro-compounds when the lone pair electrons of the basic nitrogen can play the part of Y$^{\ominus}$ in the transformations of Chart 3.5.

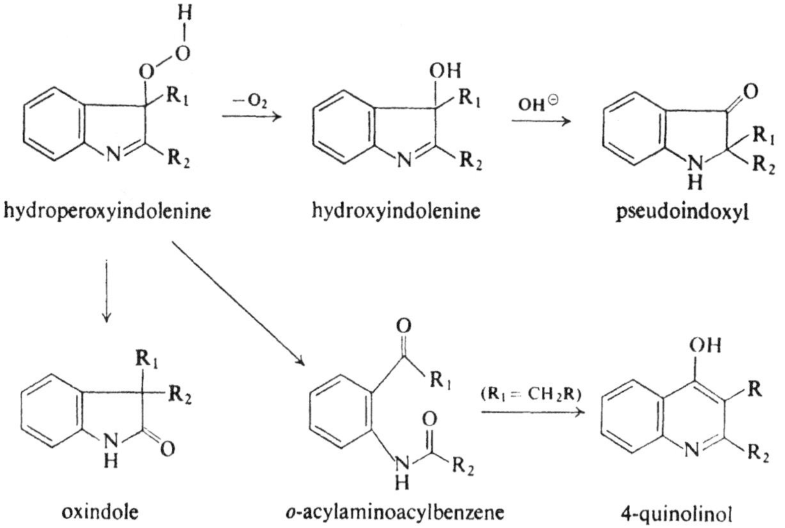

5. It is known that autoöxidation of 2,3-diethylindole or 2,3-hexamethyleneindole can yield the corresponding 2-acylindole, a system which has been recognized recently in a number of natural products. The exact mechanism of the reaction is not yet known but following Chart 3.5, the route below seems reasonable (see also p. 66).

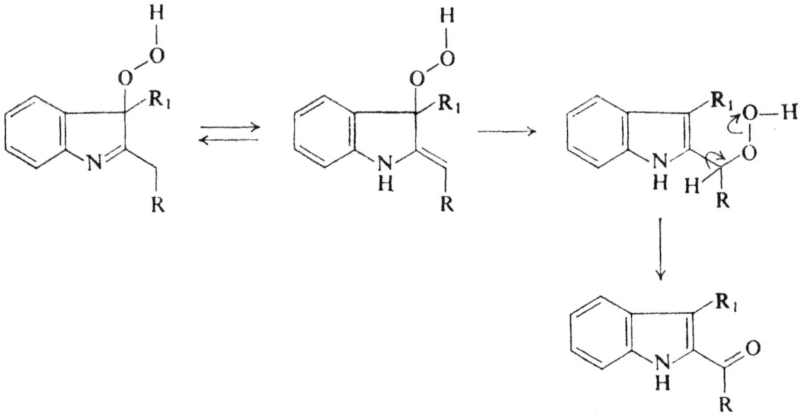

Indoline alkaloids have not been mentioned here because the key which unlocks their nuclei for degradative or interconversion work is the oxidant which converts them to the corresponding indolenine. From a biochemical point of view such an oxidation is probably the reversal of a reduction step which in essence had served to stabilize the molecule. The resultant indolenines are usually highly reactive, see for example decomposition of akuammicine Chart 3.2 and the unfolding of the hexacyclic ajmaline molecule to a tricyclic derivative (Chapter VIII).

CHAPTER IV

SIMPLE DERIVATIVES OF TRYPTOPHAN

UNDER this heading are discussed representative alkaloids derived from tryptophan equivalents in the following ways, modification of the side chain, alkylation, ring closure, dimerization and combination with mevalonic acid. A number of these bases presented no structural problem once they were analysed and recognized to contain an indolic nucleus. Others although simple contained novel elements which were not always immediately appreciated at the time of their first isolation and study. Many of these alkaloids are active principles of plants long used as drugs by native populations.

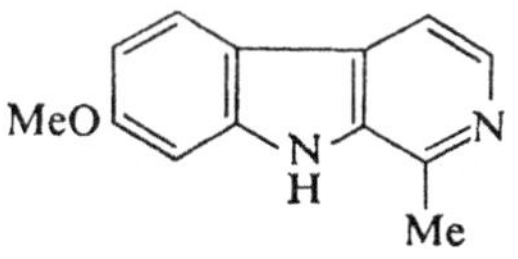

harmine

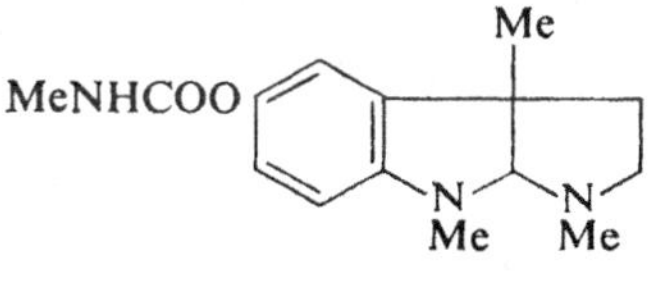

physostigmine

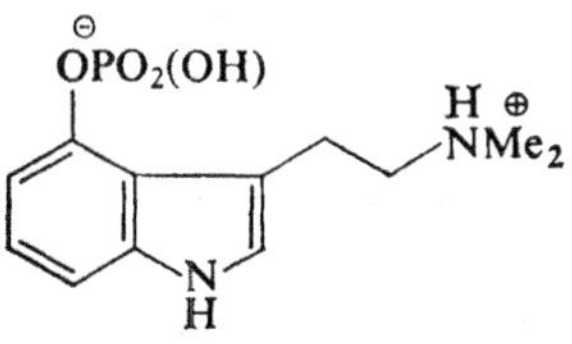

psilocybine

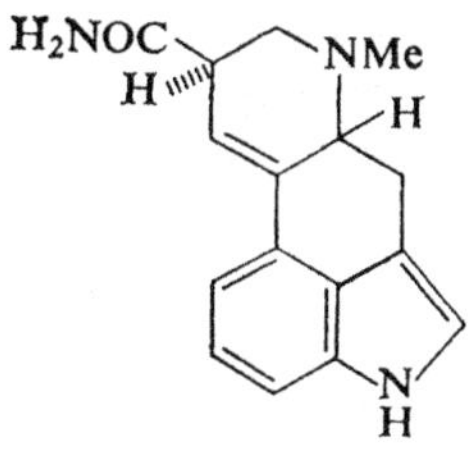

lysergic acid amide

Harmine, is for example, the main component of the narcotic drug, "yage" also called "caapi" or "ayahusca" (*Banisteria caapi*). Physostigmine (an anticholinesterase inhibitor) is contained in the calabar bean (*Physostigma venenosum*) long used in West Africa as an ordeal poison. Psilocybine from the Mexican mushroom (Psilocybe species) and lysergic acid amide from the Mexican plant "ololiuqui" (*Rivea corymbosa*) are powerful hallucinogens.

Gramine (Donaxine). This base, readily accessible via a Mannich condensation between indole, formaldehyde and dimethylamine, is produced in the plant from tryptophan by a route not completely understood but which retains the original indole moiety and the first carbon of the side chain. The usefulness of gramine to the plant is obscure but is certainly functional since in sprouting barley its concentration builds up to a maximum in ten days after germination and has practically disappeared in a month.

The most important indole in this molecule weight range is of course the auxin, indole-3-acetic acid. Parenthetically indole itself is a natural product found *inter alia* in jasmine oil and partially responsible for the odour of the flowers; woad and tyrian purple are indigo derivatives; indole acetaldehyde is also of wide distribution. Whether or not gramine plays a part in their formation is not known. Gramine is a useful intermediate in the synthesis of tryptamine derivatives.

The Tryptamines. This class of compound is of widespread occurrence amongst plants and animals some of them playing an important physiological role. Serotonine (enteramine, 5-hydroxytryptamine) has been detected in many animal sources as well as in the stinging nettle and the banana. It is a powerful smooth muscle stimulator and the closely related melatonin (N-acetyl-5-methoxytryptamine), a constituent of the pineal gland, has potent melanophore stimulating properties. The major synthetic pathways for this class of substance are shown in Chart 4.1.

CHART 4.1

SYNTHETIC ROUTES FOR TRYPTAMINES

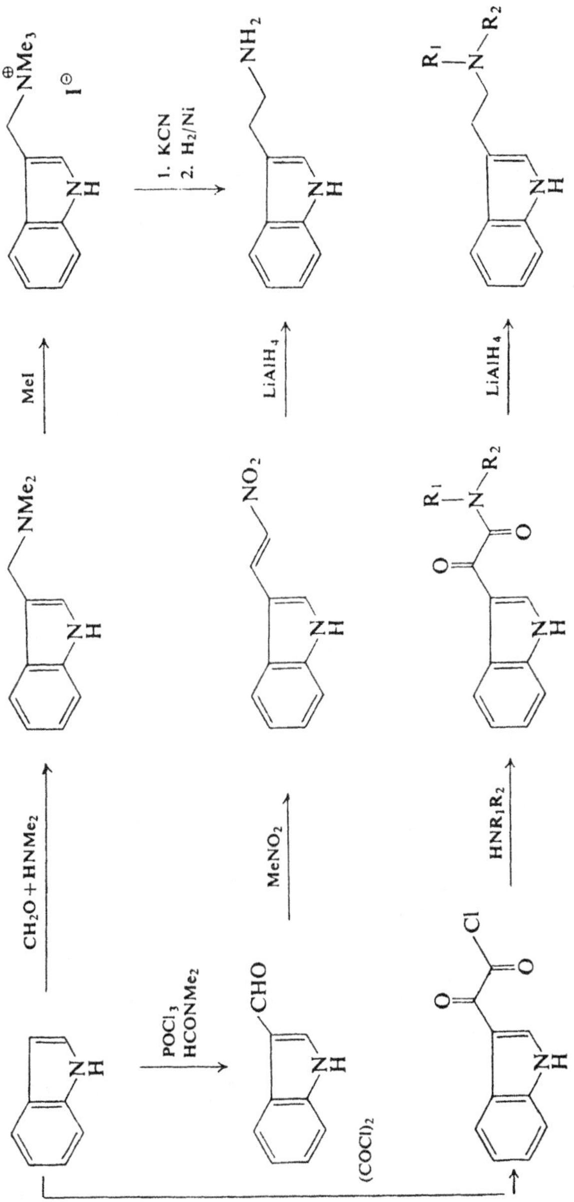

CHART 4.1—*cont.*

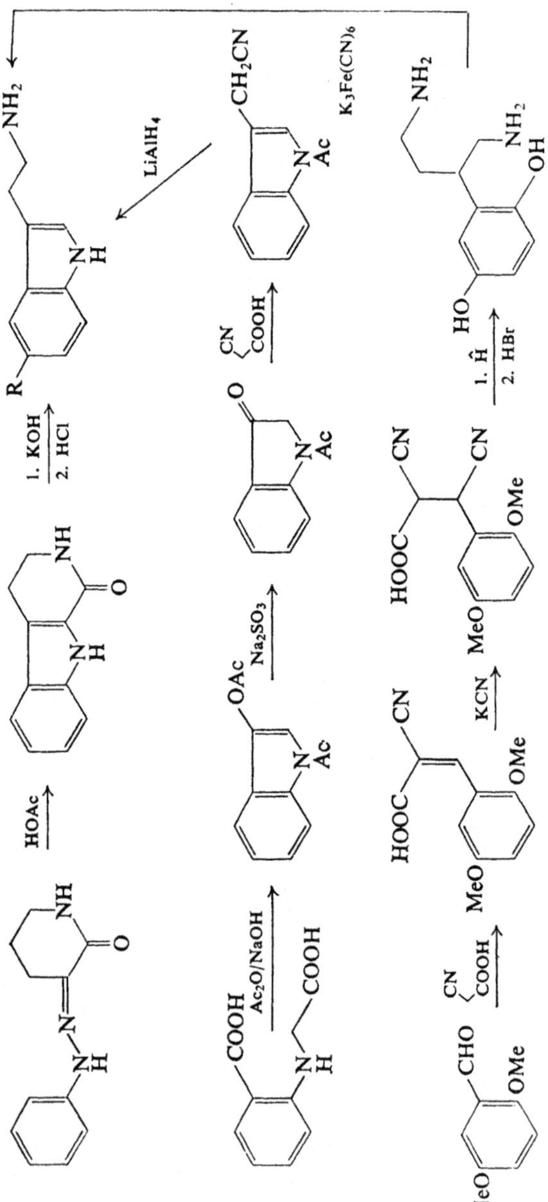

Physostigmine. This alkaloid of the calabar bean [*Physostigma venenosum*, (*Leguminosae*)] occurs along with its N-oxide, geneserine. Its use as a stimulator of the parasympathetic nervous system (because of its cholinesterase inhibiting properties) has largely been superseded by cheaper and better synthetics. Physostigmine is the methylcarbamate of the phenolic base eseroline which can be regarded as a product of trimethylation of 5-hydroxytryptamine. Experimentally, however, treatment of 5-ethoxytryptamine with ethyl magnesium iodide followed by reflux with methyl iodide gave instead the methyl derivative of the ring open tautomer (see p. 34).

The key steps in practical routes to eseroline derivatives depend in the end on cyclization of a substituted 3-β-aminoethyloxindole (p. 35).

From a degradative point of view the reader will already have guessed that ring C of eseroline derivatives would open under the proper conditions, i.e. a reversal of the original nucleophilic addition of the basic nitrogen to an indolenine or indoleninium group. Thus when a "Hofmann degradation" is applied to eserethole methohydroxide the product is the tautomeric eserethole "methine" its structure being secured by conversion to the oxindole and elimination of the terminal nitrogen. Likewise eserethole itself upon treatment with zinc and hydrochloric acid generated dihydroeserethole. Pyrolysis of the methiodide at 200° in a carbon dioxide atmosphere resulted in the complete loss of ring C and the obtention of physostigmol ethyl ether (p. 36).

The BC rings of physostigmine are probably *cis* fused (cf. chimonanthine, p. 46) since treatment of eserethole methine with acid regenerated eserethole methiodide; this means that the solution to the absolute stereochemistry will be given once the configuration of any one of the asymmetric atoms is known.

A different mode of ring closure of a tryptamine, of an oxidative-type is illustrated by dehydrobufotenine the principal product of the parotid glands of the toad (*Bufo vulgaris*). The name was originally given to the base before its structure was known but with the knowledge that it contained two hydrogens less than the co-occurring bufotenine.

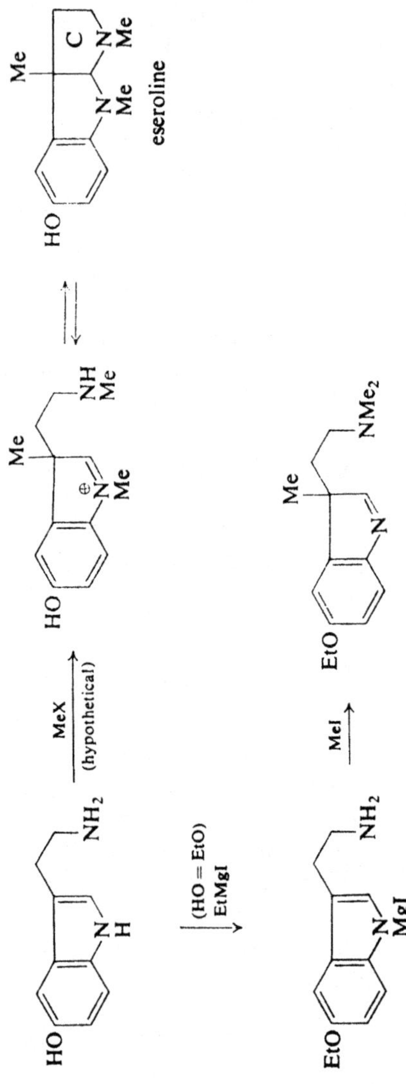
HO
NH₂
N
H
MeX
(hypothetical)
Me
NH
Me
Me
eseroline
C
(HO = EtO)
EtMgI
EtO
MgI
MeI
NMe₂

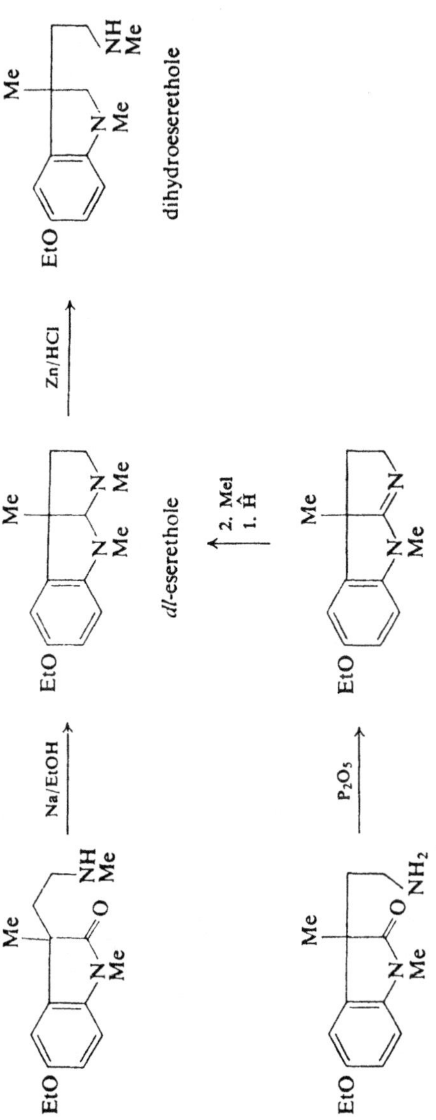
Na/EtOH
P2O5
Zn/HCl
1. H
2. MeI
dl-eserethole
dihydroeserethole

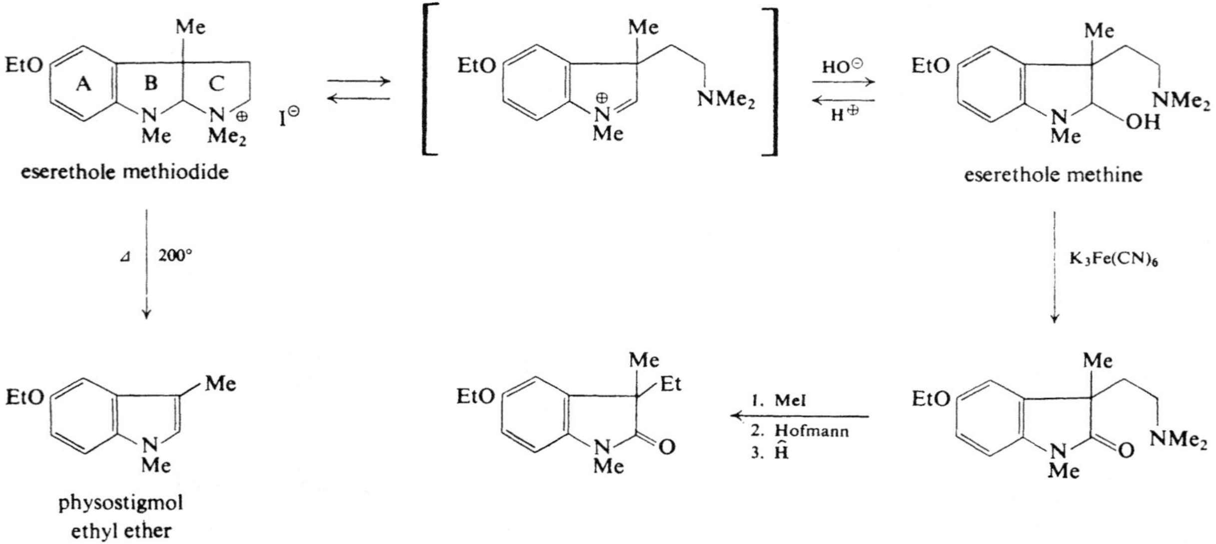
EtO
Me
A
B
C
N
Me
N
Me2
I⊖
eserethole methiodide
NMe2
HO⊖
H⊕
OH
eserethole methine
Δ
200°
K3Fe(CN)6
physostigmol
ethyl ether
Et
O
1. MeI
2. Hofmann
3. Ĥ

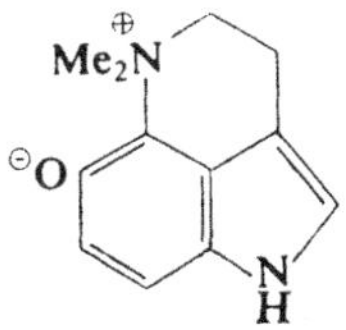

dehydrobufotenine

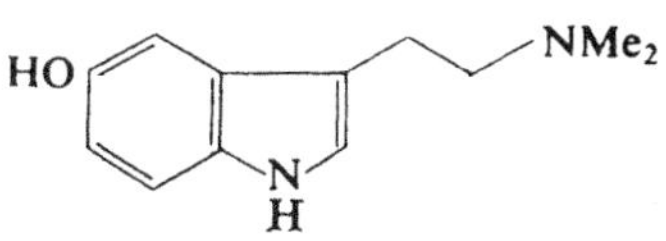

bufotenine

The Harmala Bases. Harmine (the main base), harmaline and harmalol, along with the quinazoline, vasicine, make up about 2–3 % of the seeds of the rutaceous plant, *Peganum harmala.* These bases are also found in eight other families which include the *Leguminosae, Rubiaceae* and *Passifloraceae.*

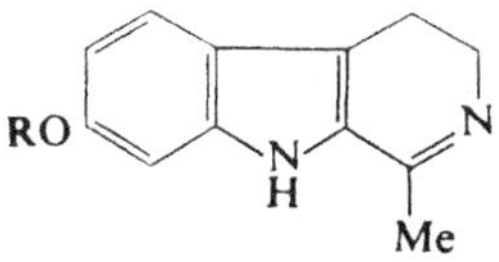

R = H harmalol
R = Me harmaline

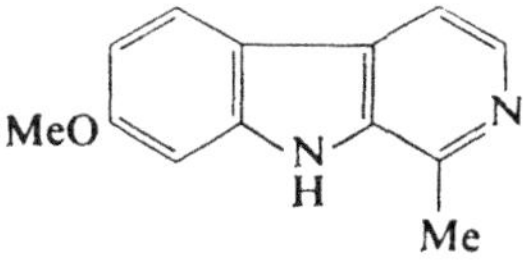

harmine

In Chart 4.2 are sketched some of the reactions whose interpretation lead to the deduction of the structures of the three bases. The work was not as simple as it is made to appear and for example the nature of the base (harman) which was originally produced by "oxidation" of tryptophan with ferric chloride was obscure until the harmala alkaloids were investigated seventeen years later and the recognition of its structure was the key to this natural products problem. This pioneering work provided a solid basis for the later development of the chemistry of the yohimbines.

Some other reactions of the harmala alkaloids are simply consequences of imine ⇄ enamine tautomerism but are nevertheless important for an understanding of the chemistry of the more complex indole alkaloids (see p. 39).

A number of synthetic methods have been developed, the most useful being extensions of those used to prepare isoquinolines. The Bischler–Napieralski condensation is carried out on the

CHART 4.2
INTERRELATIONSHIPS BETWEEN HARMALA ALKALOIDS

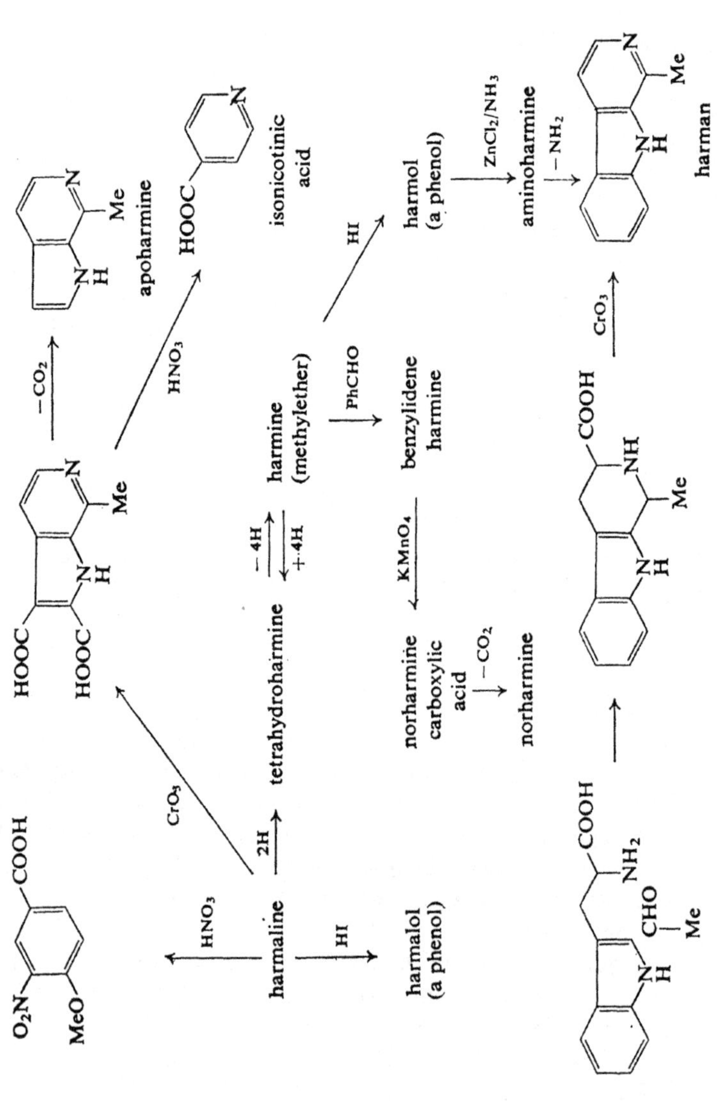

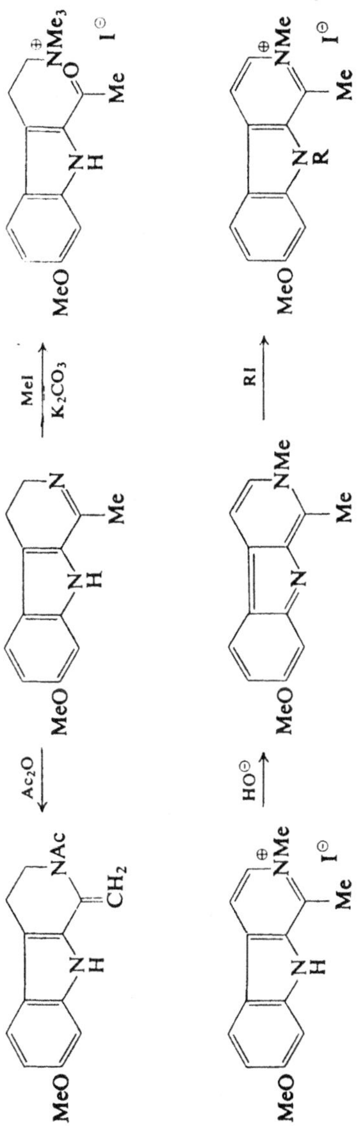

harmine methiodide

anhydro-compound

Chart 4.3
Properties of rutaecarpine and evodiamine

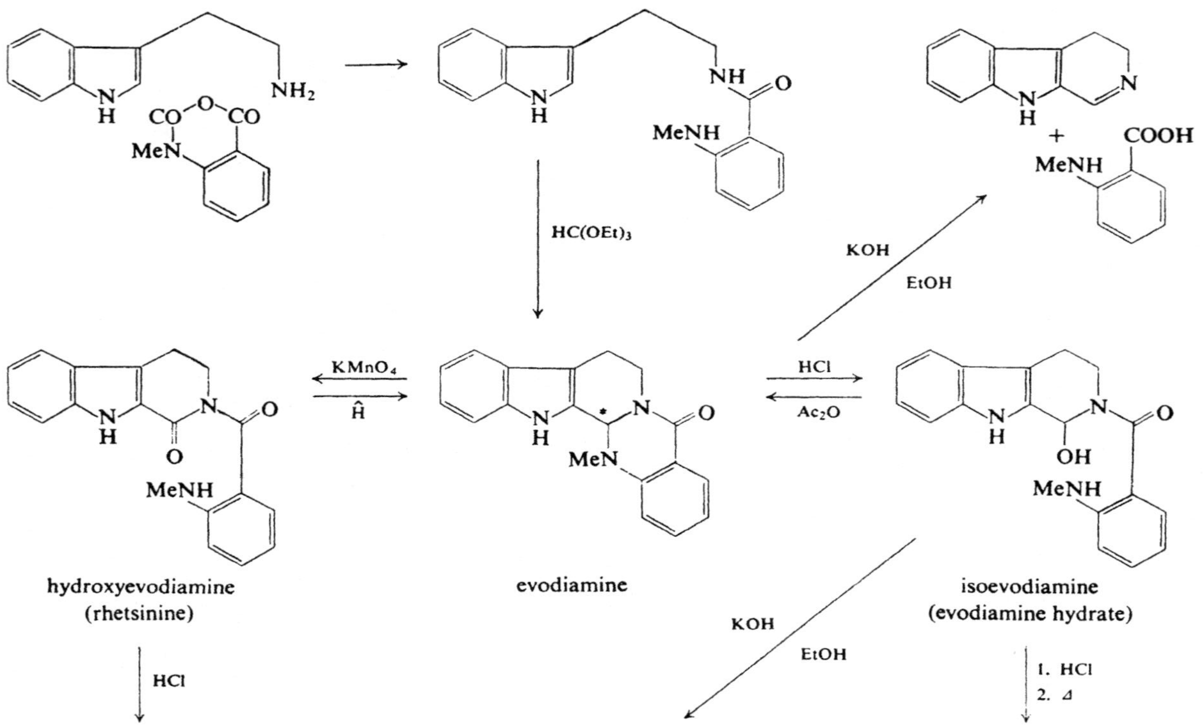

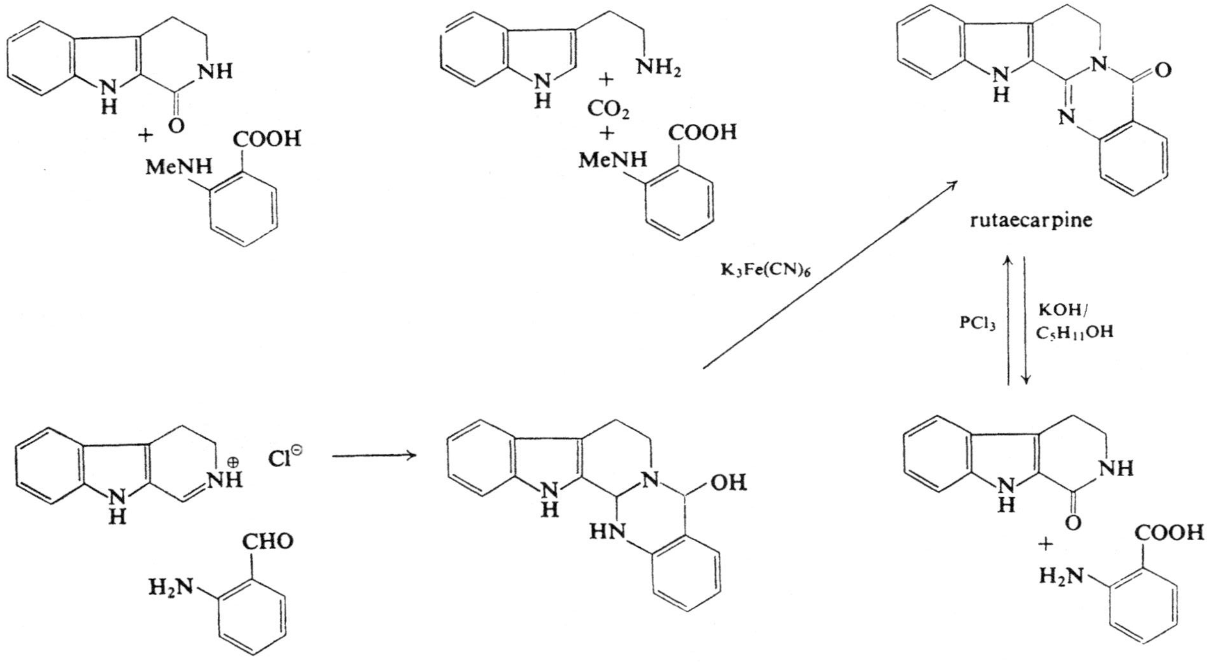

rutaecarpine
K3Fe(CN)6
PCl3
KOH/
C5H11OH

N-acyltryptamine if the harmaline system is desired and the Pictet–Spengler route, the tryptamine plus the appropriate aldehyde for a tetrahydro-β-carboline (cf. p. 25). If a β-carboline itself is required a tryptophan is substituted for the tryptamine in the latter synthesis and the product oxidized with chromic acid.

Rutaecarpine. Rutaecarpine and its optically active N-methyldihydro derivative (evodiamine) occur in the dried fruit of the Chinese drug "Wou-chou-yu" (*Evodia rutaecarpa*). Although they are quinazolones they are considered here because their chemistry is largely that of N-acyl-β-carbolines. It should be noted that the isolation procedure developed for these alkaloids does not depend on an acid extraction which is fortunate because they are weak bases. *d*-Evodiamine when heated in alcoholic hydrochloric acid furnished a hydrated product (ring fission), isoevodiamine, which ring closed again with acetic anhydride (nucleophilic addition to an acyliminium salt) to yield *dl*-evodiamine. Pyrolysis of isoevodiamine hydrochloride evolved water, methyl chloride and by oxidation or disproportionation yielded rutaecarpine. Oxidation of evodiamine with potassium permanganate in acetone gave hydroxyevodiamine (a misnomer since it is a *bis*-amide). Boiling these derivatives in alcoholic potassium hydroxide furnished β-carboline, tryptamine and anthranilic acid derivatives. A full understanding of these results leads to the structures and interrelationships shown in Chart 4.3.

The synthetic approaches are for the large part reversals of the degradative steps but in an interesting variant *o*-aminobenzaldehyde was first coupled to 3,4-dihydro-β-carboline then oxidized. *o*-Aminobenzaldehyde or its equivalent is also involved in the synthesis of other alkaloids, e.g. dictamine, vasicine and cryptolepine.

The latter alkaloid is a violet base obtained from *Cryptolepis* species (*Asclepiadaceae*) which with selenium gave quindoline which had already been prepared 45 years before. An efficient biogenetic type synthesis of cryptolepine was from indoxyl and *o*-methylamino benzaldehyde in weakly acidic medium.

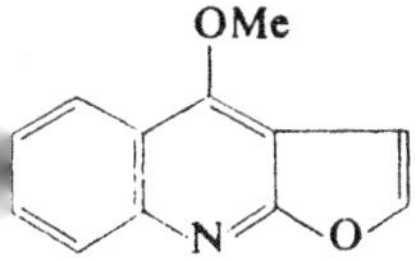

dictamine

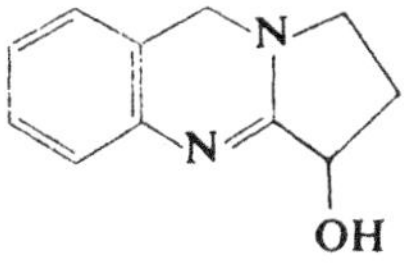

vasicine

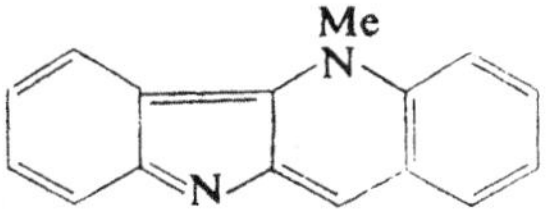

cryptolepine

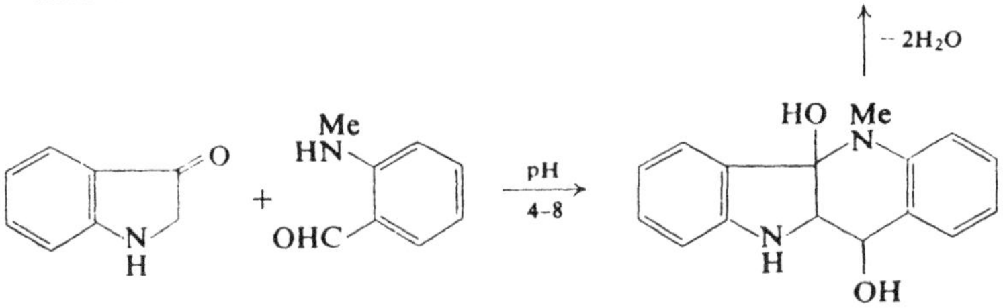

Calycanthine. A number of calycanthus species contain in their seeds quite high amounts of alkaloids which are products of the oxidative coupling of two tryptamine residues. Calycanthine, $C_{22}H_{26}N_4$, is the most well known of these compounds. Once its molecular weight was settled it was recognized to contain two secondary nitrogens (gave a dinitroso derivative) and two methyl imino groupings. Upon benzoylation and subsequent oxidation with potassium permanganate N-benzoyl-N-methyltryptamine was obtained. Upon dehydrogenation with selenium, distillation with zinc dust, soda lime and oxidation with silver acetate, the alkaloid yielded the products shown in Chart 4.4. It was the eventual recognition of the structure of calycanine, $C_{16}H_{10}N_2$, that set the stage for an understanding of the structure of calycanthine itself. Calycanine would appear to be the result of extrusion of two —$CH_2.CH_2.NCH_3$ moieties and 2 hydrogens from calycanthine. If the genesis of calycanthine is via oxidative coupling at the β positions then its structure is that possibility in Chart 4.5 which best fits the experimental data. Calycanthine is that isomer which is optically active and which can generate the fission products in Chart 4.4. It is also a thermodynamically stable isomer since it was unchanged upon reflux in acidic solution. It has been hypothesized that calycanthine might be the end product of acid catalysed equilibrations of these isomers, however, chimonanthine another natural isomer also does not isomerize in acidic medium. Two other bases calycanthidine and folicanthine are respectively mono- and dimethyl derivatives of chimonanthine. Upon oxidation with mercuric acetate calycanthine afforded an enamine rather than the sterically impossible amidine. This enamine upon hydrolysis and disproportionation gave the expected hydroxylactam and methylamine (facing page).

CHART 4.4
DEGRADATION PRODUCTS OF CALYCANTHINE

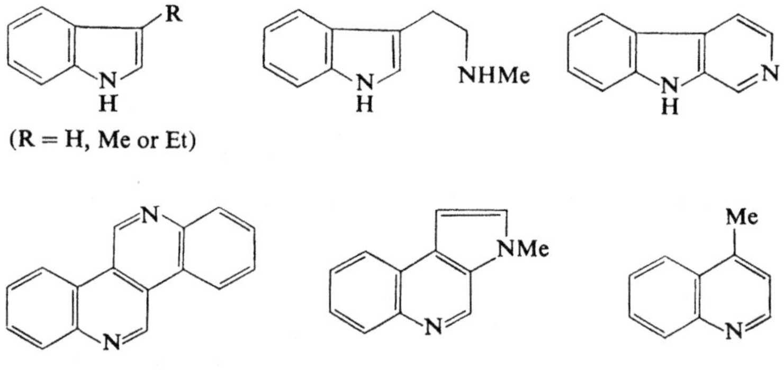

calycanine

N-methyltryptamine (dipterine) has been converted into its Grignard derivative and oxidized *in situ* with ethereal ferric chloric to yield *dl*- along with a trace of *meso*-chimonanthine. These were the expected products since they were produced under conditions where no possibility for hydrolytic ring opening was provided. In a different approach, N-methoxycarbonyl 3-2′-aminoethyloxindole was dimerized by means of iodine in tetrahydrofuran to furnish two isomeric dimers which were reduced separately with lithium aluminium hydride which converted the —NHCOOMe to —NHMe and the oxindole to the carbinolamine.

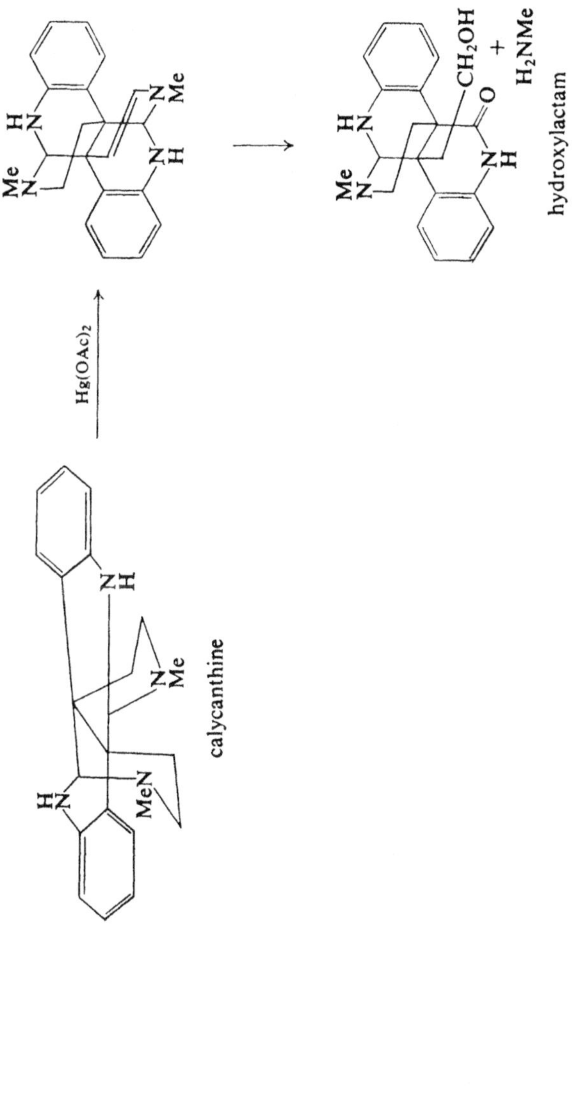

calycanthine

hydroxylactam

CHART 4.5

THE THEORETICAL OXIDATIVE DIMERIZATION PRODUCTS OF TRYPTAMINE

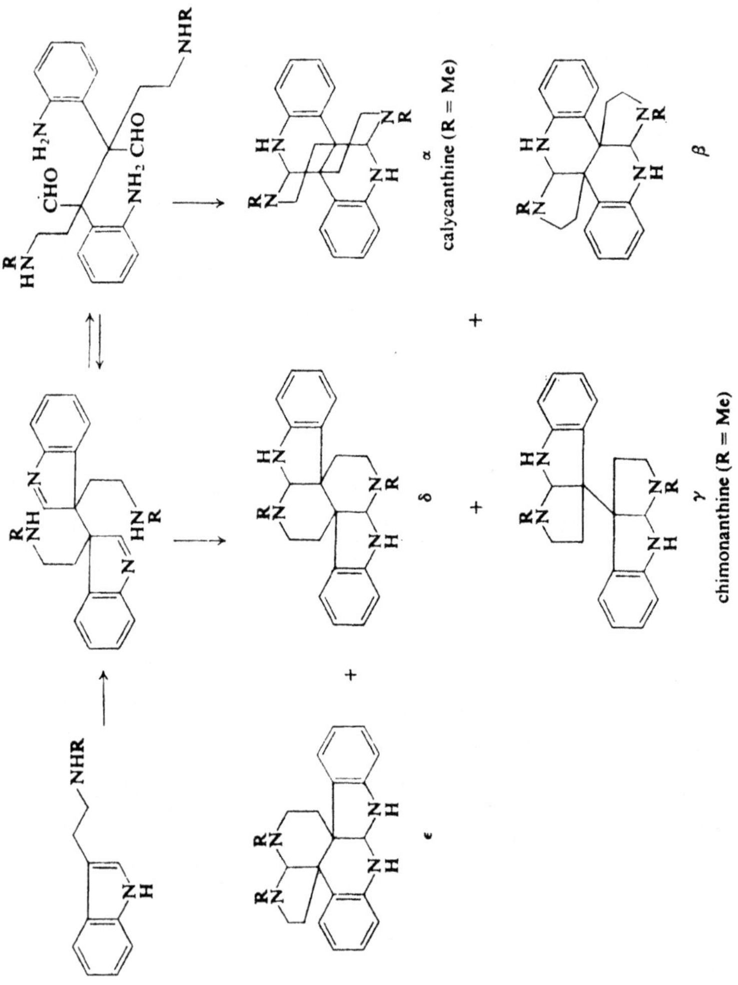

The sole isolable products were *dl*-chimonanthine from one dimer and the *meso*- form from the other.

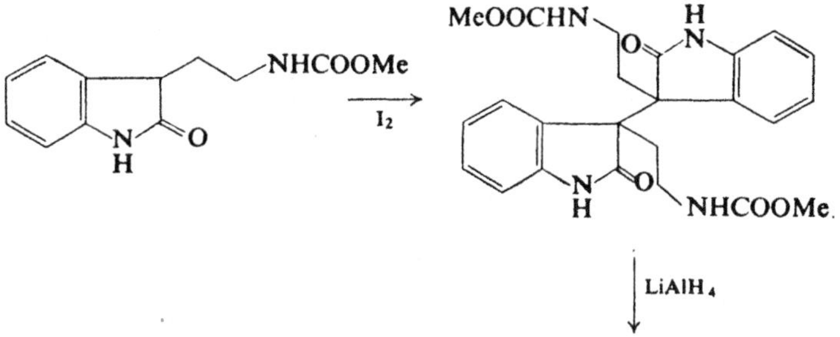

chimonanthine

Despite the unwillingness of the rings to rearrange in the case of chimonanthine the synthesis of calycanine does involve such a rearrangement which is accomplished by heating *leuco*-isoindigo in hydrochloric acid. Lithium aluminium hydride reduction and selenium dehydrogenation completed the scheme. The structures of calycanthine and chimonanthine have been derived independently of any chemical assumptions by X-ray diffraction analysis.

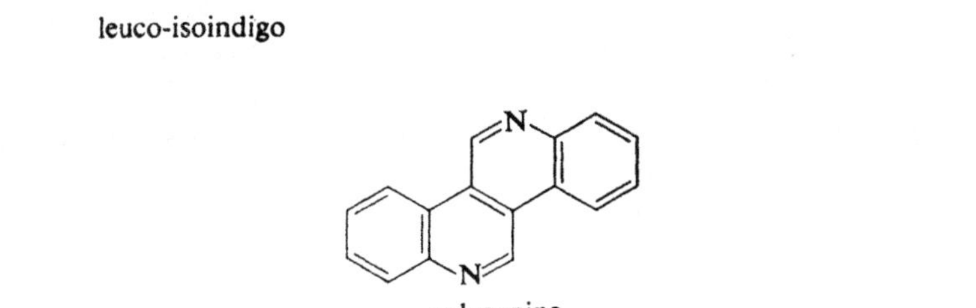

calycanine

The Ergolines. The name "ergoline" is generic for the basic heterocyclic moiety of the alkaloids of the fungus, ergot (*Claviceps purpurea*). A simple derivative of ergoline, lysergic acid amide has recently been recognized in higher plants, as the hallucinogen of the seeds of *Rivea corymbosa* (Ololuiqui) and *Ipomoea tricolor*. All the natural ergolines have a carbon substituent at C-8 and the reason for this became clear when it was found that they were biosynthesized from tryptophan and mevalonic acid. The sequence of events appears to involve attachment of an isoprenoid residue to the C-4 (C-11 of ergoline) position of tryptophan, but the later steps have yet to be elucidated.

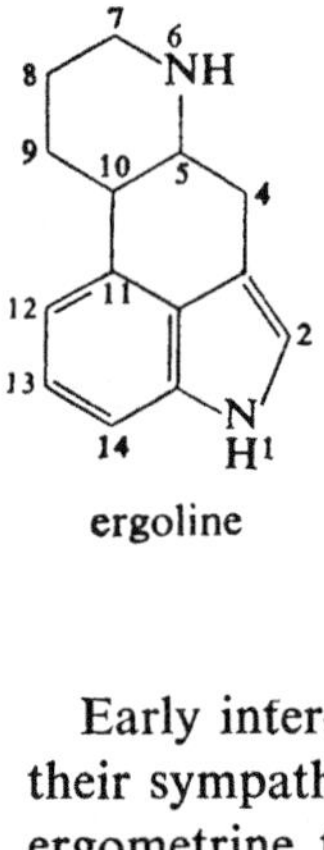

ergoline

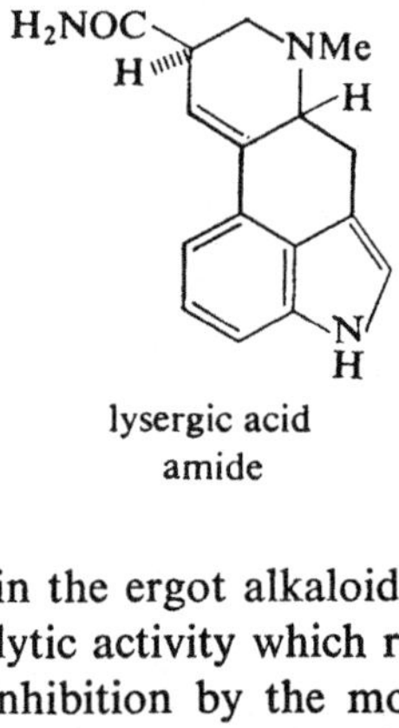

lysergic acid amide

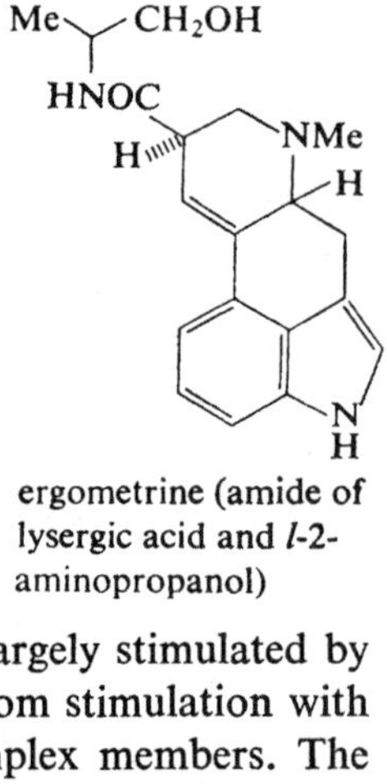

ergometrine (amide of lysergic acid and *l*-2-aminopropanol)

Early interest in the ergot alkaloids was largely stimulated by their sympathicolytic activity which range from stimulation with ergometrine to inhibition by the more complex members. The bulk of the lysergic acid is found in the fungal extracts along with its C-8 epimer, isolysergic acid, in combination with a peptide moiety whose structure will not be discussed here. There are six pairs of interconvertible isomers, and members of each pair tend to cocrystallize; one is *laevo*rotatory (lysergic acid + peptide; physiologically potent member) and the other is *dextro*rotatory (isolysergic acid + peptide; physiologically weak member). The *laevo*- form is convertible to the *dextro*- form by methanolic alkali whereas the reverse reaction occurs in acid, nevertheless lysergic acid is the only product derived from both isomers upon

CHART 4.6

RELATIONSHIPS BETWEEN ERGOTAMINE AND ERGOTAMININE

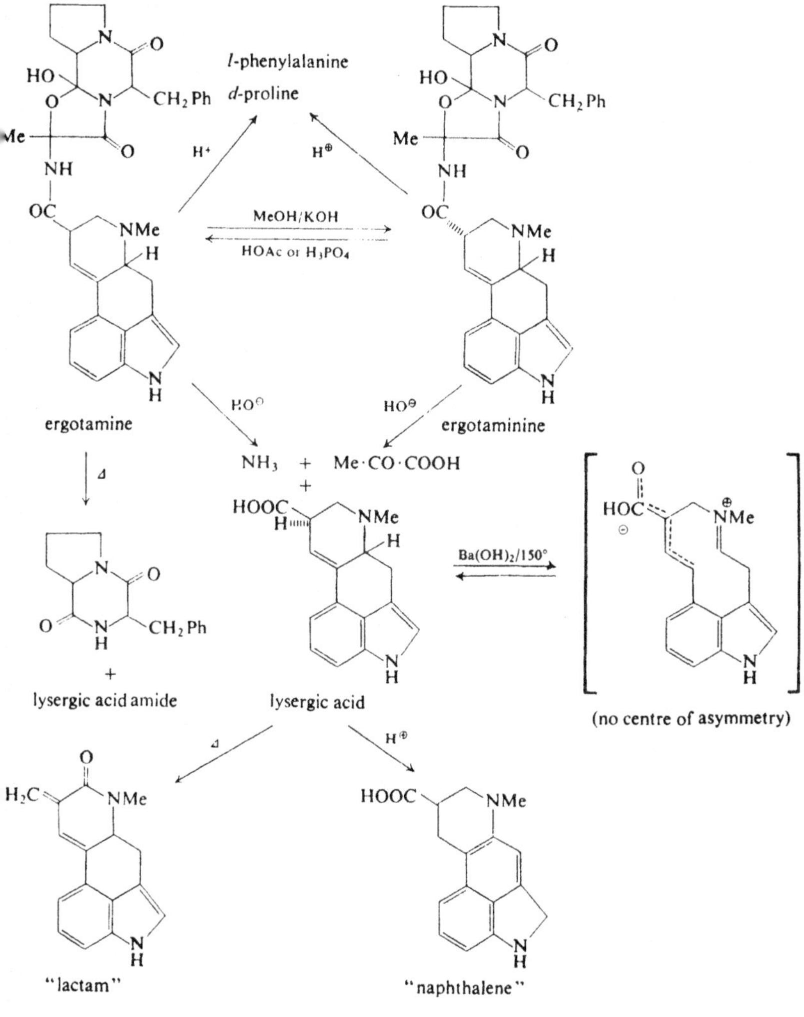

CHART 4.7
TOTAL SYNTHESIS OF *dl*-LYSERGIC ACID

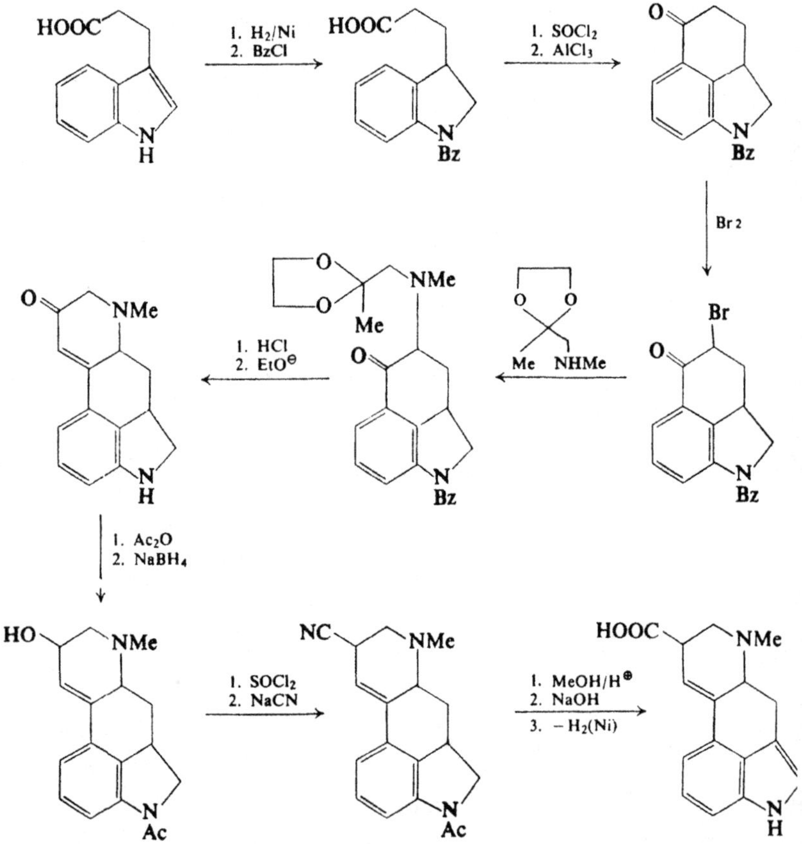

dl-lysergic acid

complete hydrolysis in base. The importance of the 9:10 double bond ($\beta\gamma$ to the carboxylic function) in facilitating the epimerization is clear and in agreement with this the corresponding dihydro compounds are unaffected under the same reaction conditions. These facts are illustrated in Chart 4.6 for the pair ergotamine—ergotaminine (the ending "ine" implies lysergic acid moiety; "inine"—the isolysergic acid moiety).

The nucleus of lysergic acid was established as the best fit to explain data, such as: 1. The production by harsh methods of, methylamine, indoles, quinoline, and 1-methyl-5-aminonaphthalene. 2. The α-position of the indole was unsubstituted according to colour reactions. 3. The double bond was conjugated to the indole (an early use of ultraviolet absorption spectroscopy in structural chemistry) and βγ to the carboxyl group. 4. The double bond must have had at least one terminus at a bridge head because two isomers were produced upon hydrogenation. 5. The double bond must in some way facilitate the racemization of lysergic acid *d*-lysergic acid but not dihydrolysergic acid is racemized by heating it in baryta at 150° under nitrogen). This may proceed via the intermediacy of an optical inactive diunsaturated acid (see Chart 4.6). 6. If lysergic acid is pyrolysed fission typical of a β-amino acid is observed resulting in an unsaturated lactam.

All doubts as to the correctness of the structure of lysergic acid were removed by a synthesis first of all of *rac*-6,8-dimethylergoline and then *rac*-dihydrolysergic acid but the comparisons were made difficult because of the production of isomers (three asymmetric centres are involved). The early approaches to the synthesis of lysergic acid itself were frustrated by its tendency under acidic conditions to isomerize irreversibly into the isomeric "naphthalene". A synthesis was finally accomplished in a simple elegant manner in which the key concept was the introduction of the indolic double bond as the last step (Chart 4.7).

SUGGESTED READING

JACOBS, W. A. and CRAIG, L. C., The ergot alkaloids XI. Isomeric dihydrolysergic acids and the structure of lysergic acid, *J. Biol. Chem.*, **115**, 227 (1936).

KERMACK, W. O., PERKIN, Jr., W. H. and ROBINSON, R., Harmine and harmaline, Part V. The synthesis of norharman, *J. Chem. Soc.*, **119**, 1602 (1921).

PERKIN, Jr., W. H. and ROBINSON, R., Harmine and harmaline, Part III, *J. Chem. Soc.*, **115**, 933 (1919).

CHAPTER V

THE COMPLEX TETRAHYDRO-β-CARBOLINE ALKALOIDS

As shown in Chart 5.1, a wide variety of alkaloid skeleta are derivable from Type I precursor and tryptamine by further ring closures. It is worth pointing out that in so far as they have been investigated, they all have had the same stereochemistry for the carbon equivalent to C-15 of yohimbine. In this chapter, certain tetrahydro-β-carboline alkaloids, viz., the yohimbines, their ring E *seco* equivalents and ring E oxygen heterocycles will be discussed since they show much chemistry in common. Over the years most of their degradation products have been synthesized, they have been interrelated, their absolute stereochemistries derived, and some have been synthesized. For these reasons they have become valuable reference compounds.

Three aspects of their chemistry are important: 1. The stereochemistry at C-3; its determination and control. 2. The chemistry of ring E (of the substituents in the case of the *seco* alkaloids). 3. The determination of the configuration and conformation of the ring systems. The solution to these problems was greatly accelerated in the years following 1952 because of the discovery and commercial importance of the antihypertensive drug, reserpine, a derivative of 3-*epiallo*yohimbane (see Chart 5.3).

The Yohimbanes. The yohimbines (Table 5.1) take their names from their first isolated and most ubiquitous member, yohimbine. They are found in rubiaceous and apocynaceous plants. Now that their structures have been established, they could be named systematically as derivatives of the pentacyclic diazahydro-carbon,

TABLE 5.1. THE YOHIMBINES

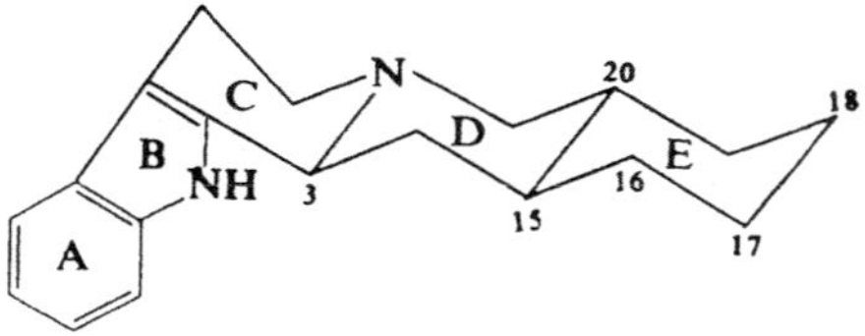

yohimbane (normal)

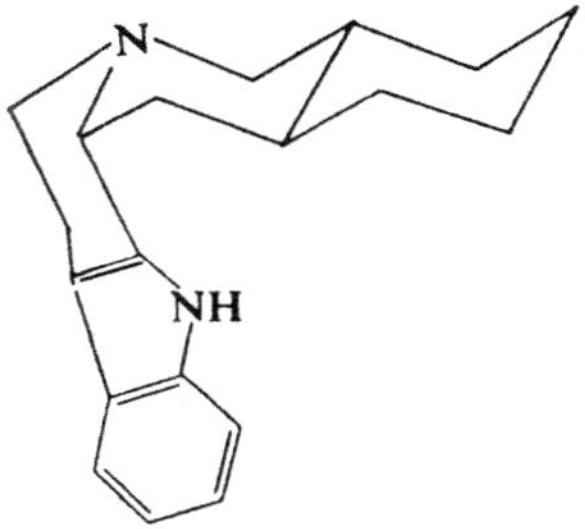

*psudoye*ohimbane

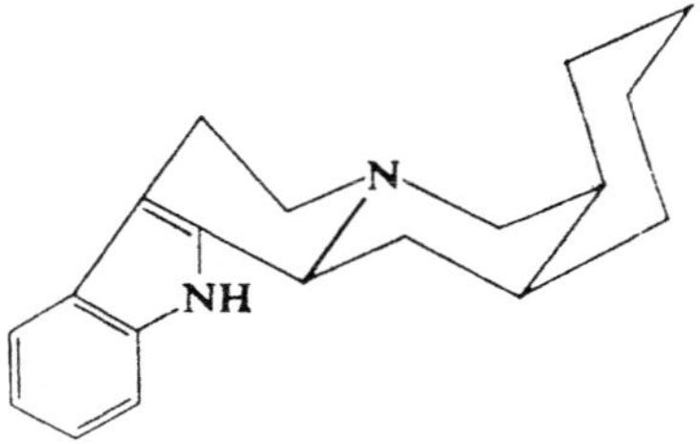

*allo*yohimbane

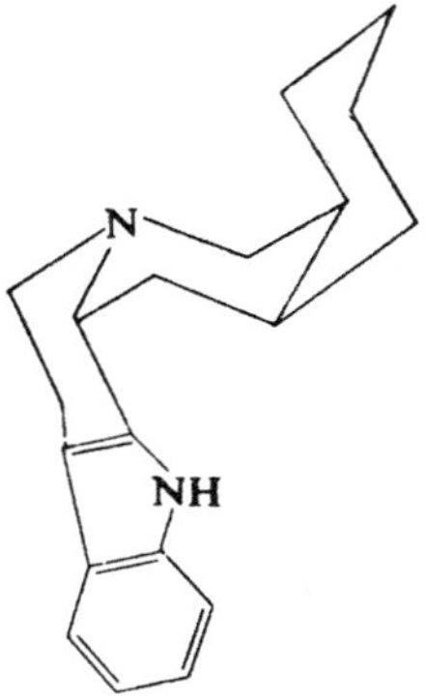

cis-trans-cis

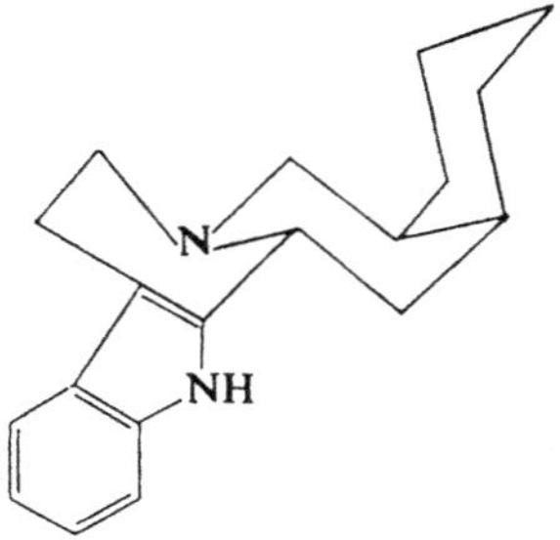

trans-trans-cis

*epiallo*yohimbane

Alkaloid	Nucleus	Configuration* —COOMe at C-16	Configuration* HO at C-17
yohimbine	normal	α	α
corynanthine	normal	β	α
β-yohimbine	normal	α	β
*pseudo*yohimbine	pseudo	α	α
*allo*yohimbine	allo	β	β
α-yohimbine	allo	β	α
3-*epi*-α-yohimbine	epiallo	β	α

* This convention is same as for steroids, viz. α denotes a substituent below the plane of the paper and β a group above, in the projection drawing.

3

CHART 5.1

BIOGENETIC RELATIONSHIPS BETWEEN SOME TYPE I ALKALOIDS

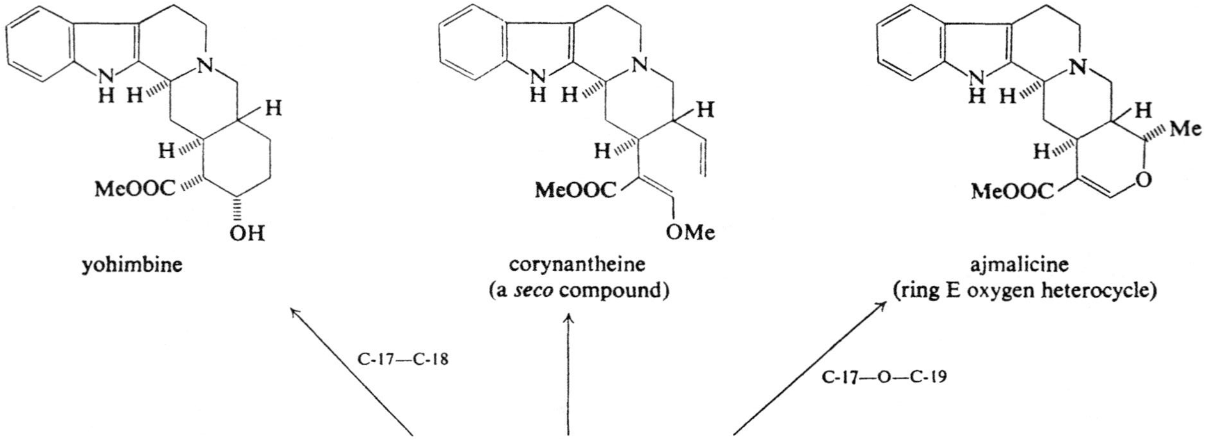

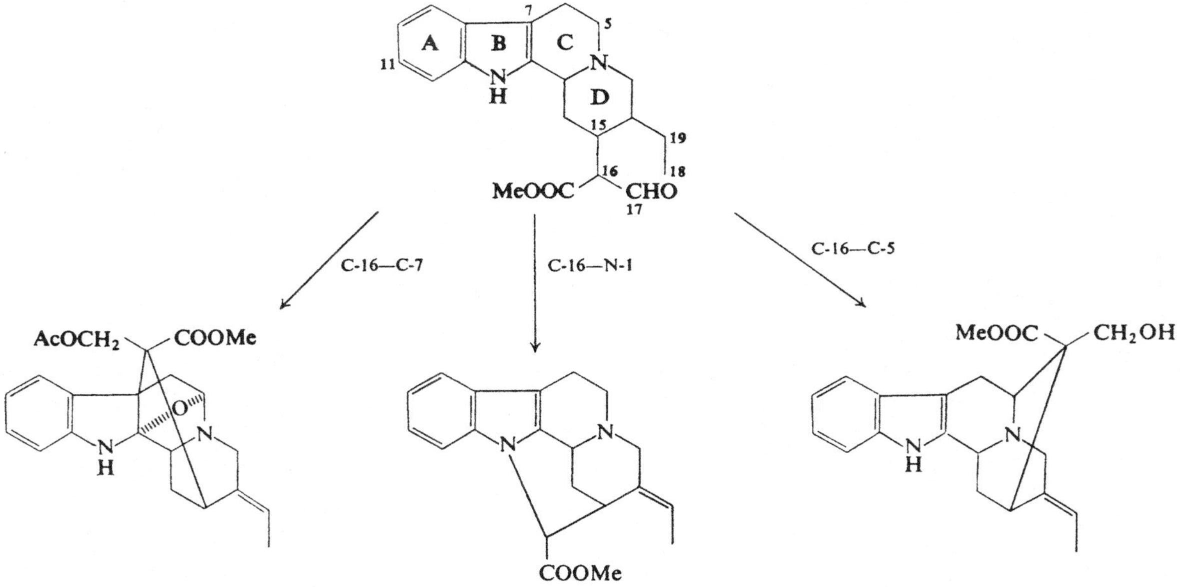

picraline pleiocarpamine akuammidine

yohimbane, rather than retaining the cryptic prefixes. Four diastereoisomeric yohimbane skeleta are found in the group, the normal yohimbane, itself, *pseudo* (3-epiyohimbane), *allo* (20-epiyohimbane, i.e. *cis*- DE ring junction) and *epiallo* (3,20-diepiyohimbane). In the latter case the two extreme conformations can be described in terms of the stereochemistry of the pairs, N-4 unfilled orbital versus C-3 proton, C-3 proton versus C-15 proton and C-15 proton versus C-20 proton, i.e. *cis-trans-cis* or *trans-trans-cis*. The complicating feature of *epiallo*yohimbane chemistry is a result of its conformational mobility, the preferred conformation being determined by steric effects of substituents which in practice means for natural products, groups at C-16, C-17 and C-18. Since there is little energy difference between *allo*- and *epiallo*yohimbanes, it is often difficult to predict the product(s) when such derivatives are subject to C-3 epimerizing conditions. It is for these reasons that determination of the axial nature of the C-3 proton by physical methods (see below) cannot by itself be diagnostic for *allo* versus *epiallo* as it can be used to distinguish between normal and *pseudo* in the *trans*- DE ring systems. These statements are illuminated by some examples:

1. Wolff–Kishner reduction of *epiallo*yohimban-17-one affords a mixture of *allo*- and *epiallo*yohimbanes in contrast to the complete conversion of *pseudo*yohimban-17-one to yohimbane.

2. Reserpine (an *epiallo*yohimbane derivative, Chart 5.3) exists largely in the *cis-trans-cis* form with its three ring E substituents equatorially oriented rather than suffer the steric crowding required of the *trans-trans-cis* conformer with the groups now all axial. Treatment of reserpine with sodium methoxide generates methyl reserpate and methyl trimethoxybenzoate but prolonging the conditions results in epimerization of the 16-methoxycarbonyl and 17-methoxyl groups and the changed steric requirements now favour the *trans-trans-cis* conformer. Under acidic conditions reserpine epimerizes at C-3 to give isoreserpine. In the course of a total synthesis of reserpine the course of the last isomerization was reversed by creation of the proper ring E derivative. Isoreserpic acid was converted into its lactone which upon reflux

in pivalic acid inverted at C-3 to give the much less hindered reserpic acid lactone (Chart 5.3).

3. Similarly to reserpic acid lactone, in rauniticine, the *epiallo* system, specifically the *trans-trans-cis* conformer, is favoured under acid equilibrating conditions because it releases the molecule from the steric crowding of the axial 19β-methyl upon the lone pair nitrogen orbital and the 14β-proton in both the *allo* form and the *cis-trans-cis epiallo* conformer.

CHART 5.2

EPIMERIZATION OF RAUNITICINE

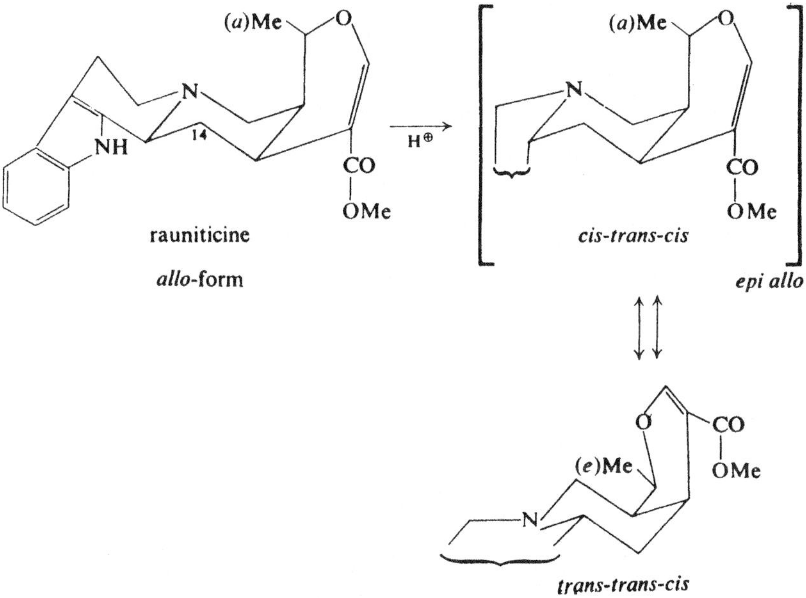

In *allo*yohimbane derivatives advantage has been taken of the ability of N-4 to quaternize internally with properly oriented ring E substituents. Thus, tosylation of 3-*epi*-α-yohimbine in pyridine yields a quaternary salt, as also do 18-hydroxyl and 16β-hydroxymethyl substituents. Such reactions yield extremely

CHART 5.3

SOME PROPERTIES OF RESERPINE (PARTIAL FORMULAE)

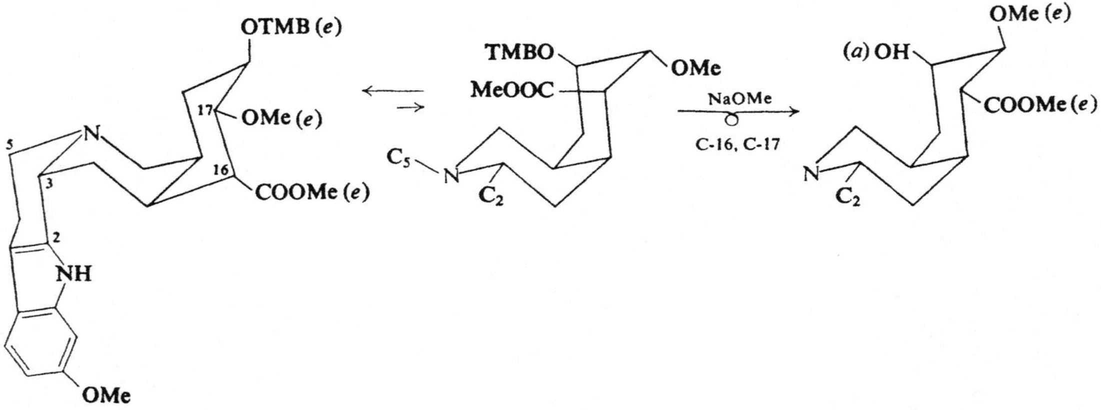

cis-trans-cis | *trans-trans-cis* | methyl neoreserpate

reserpine

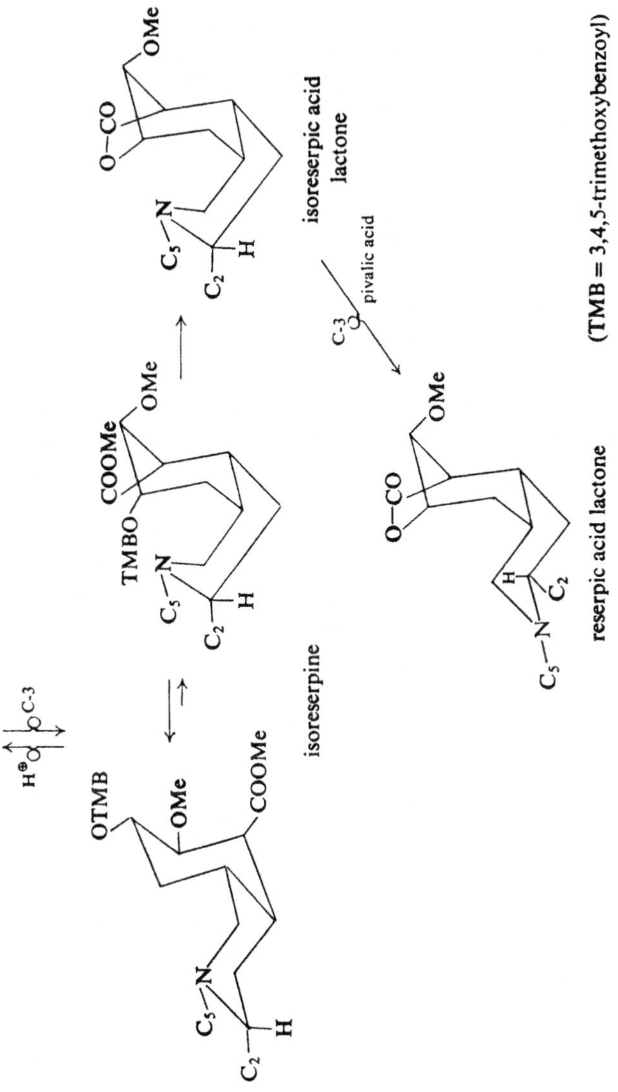
OTMB
OMe
COOMe
C_5
N
C_2
H
$H^{\oplus}$
α
β
C-3
TMBO
COOMe
OMe
isoreserpine
O—CO
OMe
isoreserpic acid lactone
C-3
α
pivalic acid
reserpic acid lactone
(TMB = 3,4,5-trimethoxybenzoyl)

useful stereo-chemical information and are diagnostic for the *cis* ring junction.

The Yohimbines. The alkaloids were originally established to be members of a set by yielding the same dehydrogenation products as yohimbine (Chart 3.1). It should be noted that the obtention of ketoyobyrine (a misnomer) fixed the site of the methoxycarbonyl in the original molecule and that the dehydrogenation products taken together define the overall structure leaving only the position of the hydroxyl and stereochemistry to be settled. In the specific case of corynanthine, it is interesting to note the formation also of *ind*-N-methylyobyrine (methyl transfer —COOMe→NH).

The chemistry of the yohimbines forms an excellent example of the value of studying related alkaloids together and Chart 5.4 which illustrates this and its obvious conclusions, can be checked against Table 5.1. The absolute stereochemistry of the yohimbines has been looked at by most of the optical comparative methods. A purely chemical proof depends on the conversion of yohimbine into a *seco* base which in turn has been related to cinchonine (*infra*).

Historically it was a total stereospecific synthesis (p. 25) of 3-*epiallo*yohimbane in 1954 that put all the conclusions up to that time on a solid basis. It also confirmed the opinion that the platinum catalysed hydrogenation of sempervirine was an all *cis* addition. Sempervirine (a constituent of yellow jasmine, *Gelsemium sempervirens*) had previously been synthesized thus.

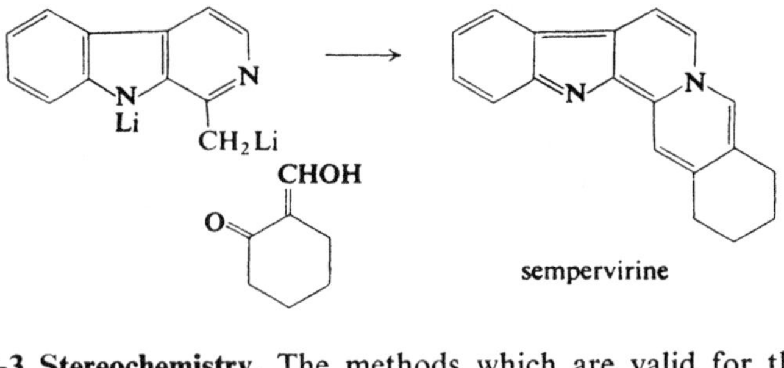

C-3 Stereochemistry. The methods which are valid for these tetrahydro-β-carboline alkaloids depend for the most part on the

CHART 5.4

CHEMISTRY OF YOHIMBINE AND ITS CLOSE RELATIVES (PARTIAL FORMULAE)

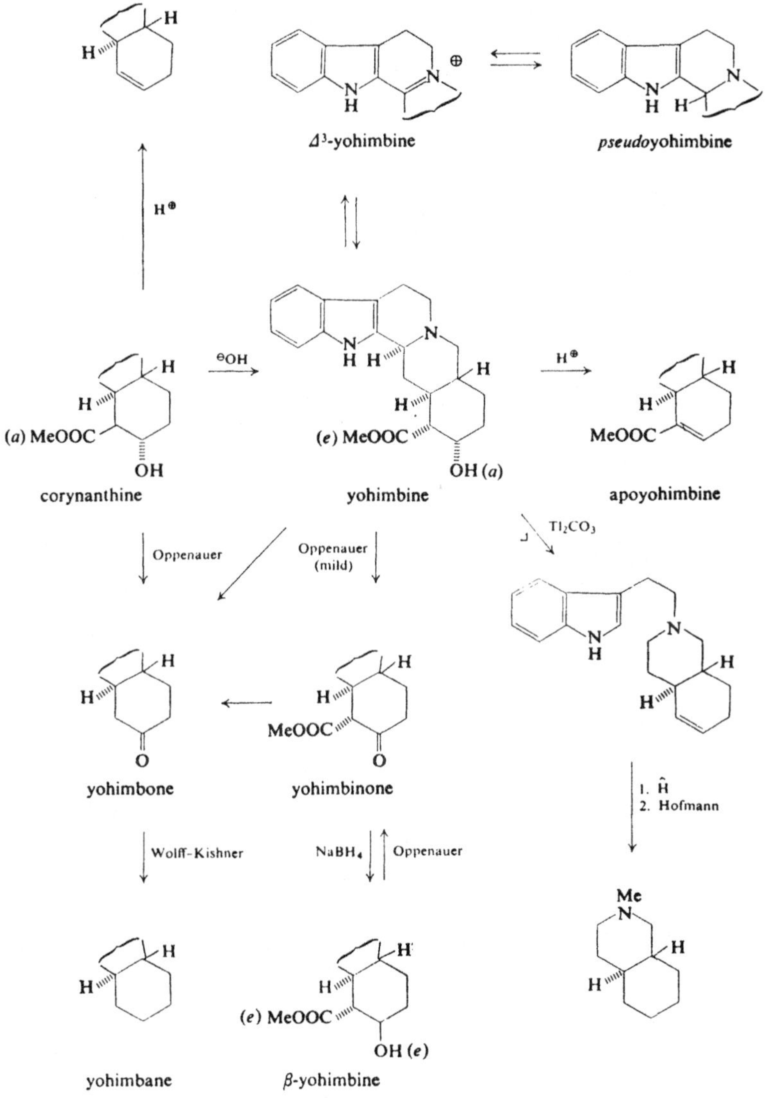

presence of the quinolizidine moiety (rings C and D). Where the 3-hydrogen is axial, i.e. α and part of a *trans*-quinolizidine moiety it is readily oxidized by mercuric acetate to a Δ^3-derivative (Chart 3.4), the *cis*-isomer is stable under the conditions. *trans*-Quinolizidines show bands (Bohlmann bands) in the infrared *ca.* 3·4–3·7μ which were originally thought to be diagnostic for the 3α-hydrogen before it was realized that the C-3 proton of a *trans-trans-cis epiallo*yohimbane also satisfied the same criterion (Table 5.1). The proton at C-3 sitting between the indole and N-4 is often moved far enough downfield to be clearly observed in the nuclear magnetic resonance spectrum. With a sufficient number of model cases on hand, it should be possible to state whether it is axially or equatorially oriented. The chemical shift of the N-methyl of the corresponding quaternary salts is of diagnostic value as also is the rate of alkylation a function of the overall stereochemistry.

Chemical methods depend on acid catalysed equilibrium studies and hydrogenation of the derived tetradehydro and/or Δ^3-compound to get with say, sodium borohydride predominantly the thermodynamically more stable (usually 3α isomer) or with zinc and perchloric acid the less stable usually the 3β isomer. As has been stated before with *allo* systems care must be taken in the interpretation of any single experiment.

The *seco* Compounds. The structures of these compounds were rapidly elucidated once systematic studies were undertaken (Chart 5.5). Of great value was the use of the Kuhn–Roth oxidation as a simple method to monitor the presence or absence of a C-methyl group. The properties of ring C and its 3α proton parallel those of the same moiety in yohimbine, and as might be expected selenium dehydrogenation furnished the tetrabyrine analogue, alstyrine.

The presence of the β-methoxyacrylate residue was indicated by the ultraviolet and infrared absorption spectra and followed from the reactions sketched in Chart 5.5. Dihydrocorynantheine and corynantheidine (20-epi-dihydrocorynantheine) underwent equivalent transformations. The depicted orientation of the β-methoxy-

CHART 5.5

TRANSFORMATIONS OF *seco* ALKALOIDS ILLUSTRATED BY CORYNANTHEINE

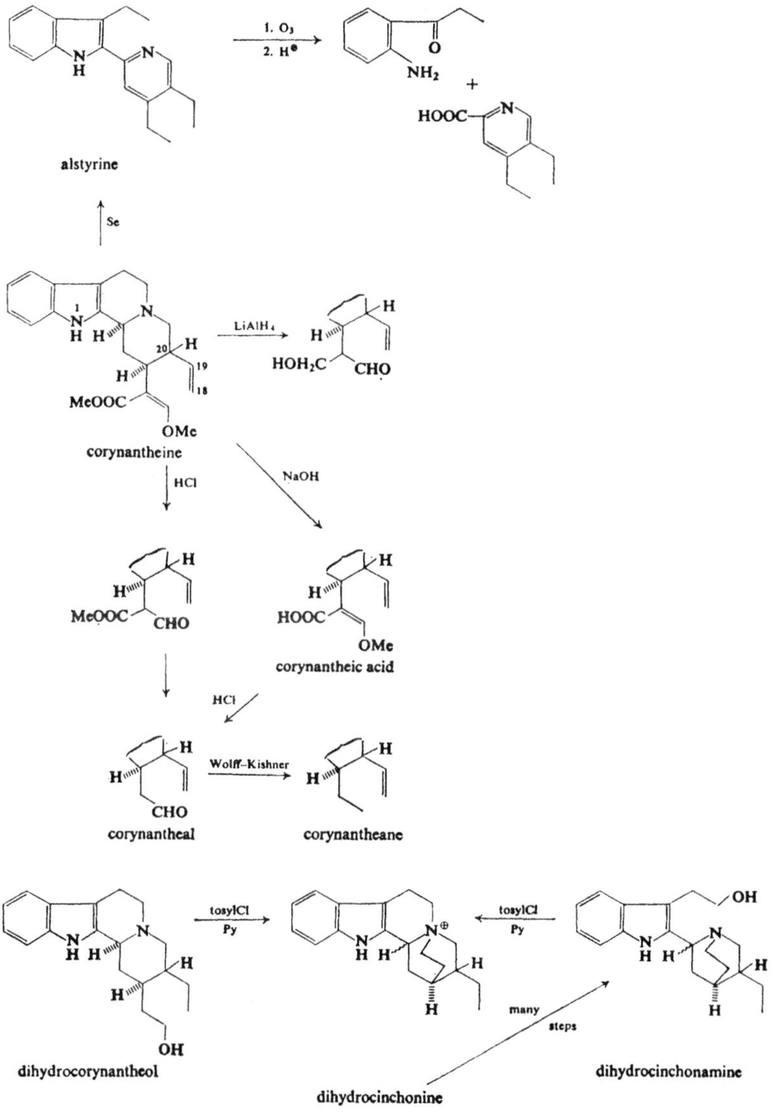

acrylic ester moiety is in accord with the interpretation of the nuclear magnetic resonance spectra.

The absolute stereochemistry of the group depended on a number of arguments. For C-3 it is based on a comparison of the molecular rotation differences between the bases and their dehydro and tetradehydro derivatives with analogous yohimbine derivatives. For C-15 on the conversion of dihydrocorynantheol and dihydrocinchonamine into the same quaternary salt. Dihydrocinchonamine had previously been prepared from cinchonine by a route which did not affect the stereochemistry at C-15.

Burnamicine. This natural constituent isolated in trace quantities from *Hunteria eburnea* is included here because of its obvious relationship to corynantheol and because it represents a good example of the use of mass spectrometry in structure deduction. Burnamicine had an ultraviolet absorption maximum typical of a 2-acylindole which changed to that of a typical indole upon either sodium borohydride reduction or solution in acid. After the latter experiment the original chromophore was regenerated in basic solution. From the nuclear magnetic resonance spectrum the four aromatic protons, the ethylidene and the N-methyl were identifiable. The recovered alkaloid upon acetylation had an ultraviolet and nuclear magnetic resonance spectrum consistent for an *ind*-N,O-diacetate (probably primary).

The mass spectrum gave a picture completely consistent with the structure (Chart 5.6). The predicted fissions were observed and amongst others gave (with hydrogen transfer from right to left) for *a*, m/e 130 (very strong) and 196, for *b*, 158 and 168 (very strong) and *c*, 172 and 154. Strong peaks at m/e 168, 130 and 144 (strongest peak in the spectrum) are those which are consistent with the extrusion of carbon monoxide (m/e 28) from fragments m/e 196, 158 and 172. The hydrogen transfer mentioned above is characteristic of carbonyl containing compounds.

A synthesis of dihydroburnamicine has been accomplished by re-arranging 7-acetoxy-7H-dihydrocorynantheol methosalt in aqueous acetic acid (cf. general principles outlined in Chart 3.5).

CHART 5.6
CHEMISTRY OF BURNAMICINE

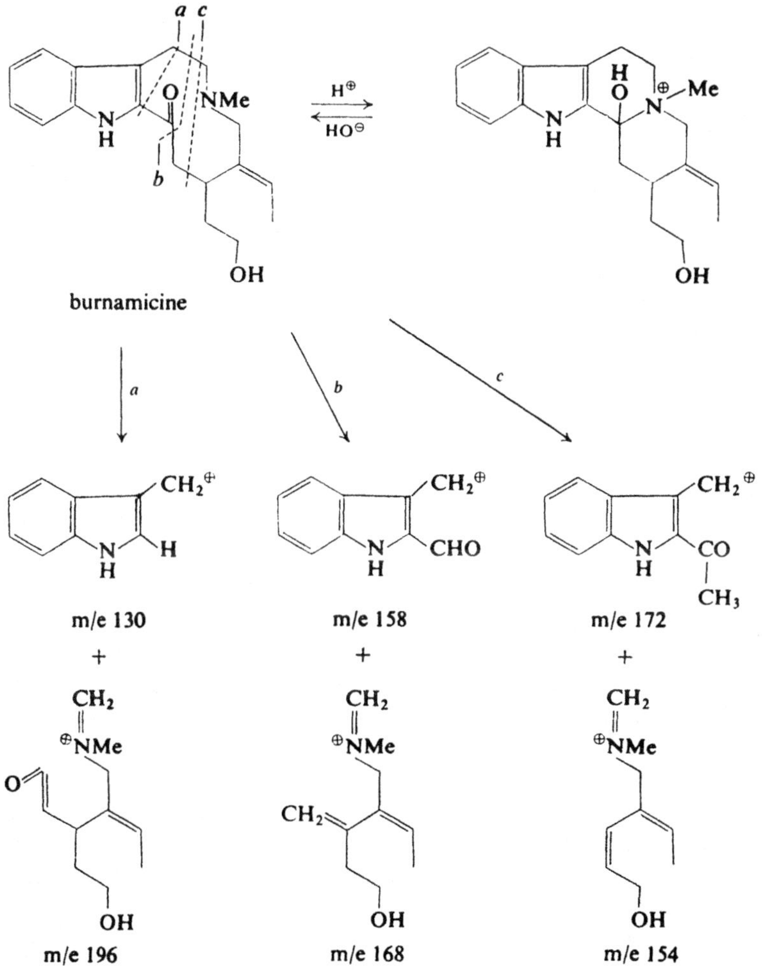

Some subtle factors must be involved in the success of this reaction since 7-acetoxy-7H-yohimbine gave the $\Delta^{3,4}$ derivative.

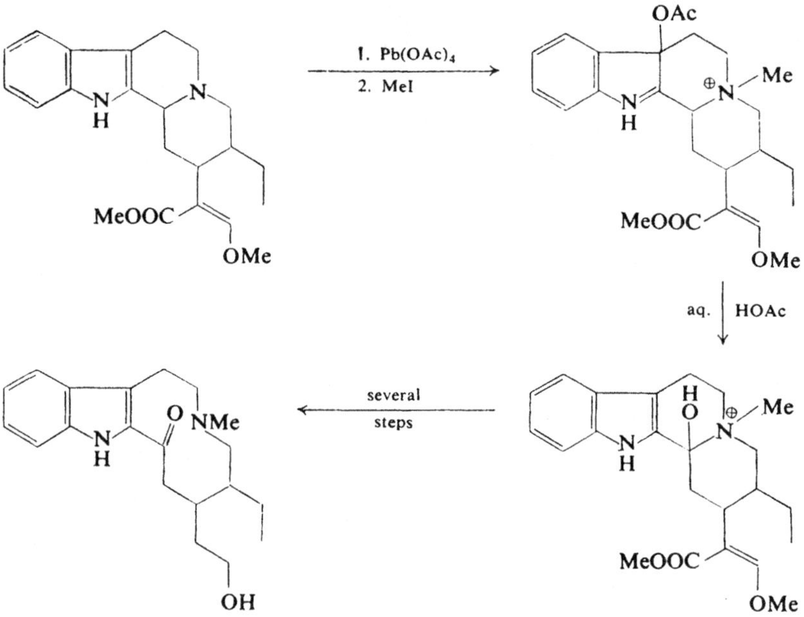

Hunterburnine α- and β-Methochlorides. Similar to the case of burnamicine, the quaternary salts hunterburnine α- and β-methochlorides were isolated from *Hunteria eburnea* in amounts which did not lend themselves to chemical examination and they were not volatile enough for mass spectrometric work. By physical methods they were recognized to be 5-hydroxyindoles, to contain a quaternary N-methyl and a vinyl group. Since their iodides were in a form suitable for X-ray analysis this appeared to be the method of choice for their structural elucidation which is discussed for the β-methiodide.

Hunterburnine β-methiodide crystallized in the orthorhombic system space group $P2_12_12_1$ (a common one for natural products) with four molecules to the unit cell. Three dimensional X-ray

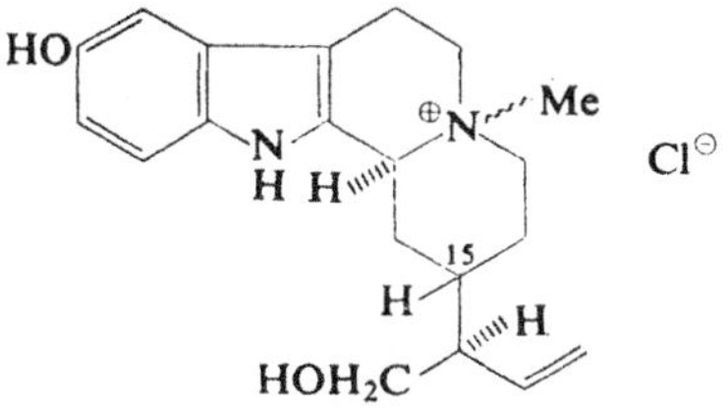

Hunterburnine α- and β- methochlorides

intensity data were recorded on equi-inclination Weissenberg photographs and in all 1784 independent amplitudes were measured.

The co-ordinates of the iodine atom were obtained initially from a Patterson synthesis then successive three dimensional Fourier syntheses were carried out with the inclusion of further atoms as they became recognized. After about five rounds of structure factor

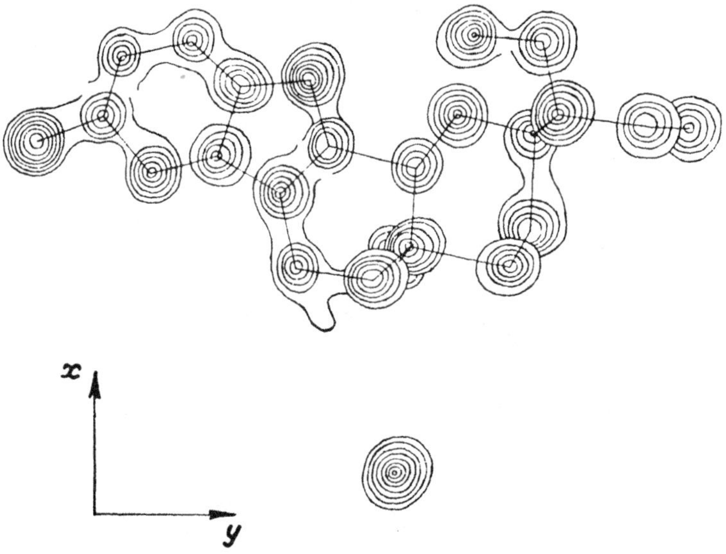

FIG. 1. The sixth three-dimensional electron-density distribution over one molecule of hunterburnine β-methiodide shown by means of superimposed contour sections drawn parallel to (001). (Reproduced by permission of The Chemical Society.)

and Fourier calculations the structure was established. Further refinement led to the illustrated three dimensional electron density distribution over one molecule of the alkaloid. The stereo chemistry at C-15 is based on the assumption that it is the same as the equivalent carbon in yohimbine.

Because of the advent of electronic computers the calculations involved no longer represent a time consuming problem with the result that the structure of a substance of this complexity containing a heavy atom is often deduced in about three months.

Hunterburnine α- and β-methochlorides have turned out not only to be representatives of a previously unexpected combination of Type I precursor and tryptamine but also to be the first examples of naturally occurring epimeric quaternary salts.

Therefore in spite of the paucity of chemical knowledge concerning them, they have raised interesting biogenetic questions. Did they originate via methylation of their tertiary precursors? Or were they the result of a ring closure (natural or during the isolation procedure) of a tricyclic burnamicine-like molecule?

Ring E Oxygen Heterocycles. The isomeric bases ajmalicine and tetrahydroalstonine (along with their tetradehydro compounds, serpentine and alstonine) are the most well known members of this group and had to be closely related to the *seco* compounds since upon dehydrogenation they yielded alstyrine. From the empirical formulae, absence of an isolated double bond, presence of C-methyl and a β-oxyacrylic ester moiety (saponification and ultraviolet absorption spectra) the structures of ajmalicine and alstonine were derived. The stereochemical problems were not completely resolved until the total synthesis of ajmalicine was accomplished.

The α-orientation of the C-3 proton in both ajmalicine and tetrahydroalstonine was established by those methods previously discussed for the other tetrahydro-β-carboline alkaloids. Ajmalicine was converted into dihydrocorynantheane by the steps indicated in Chart 5.7, which fixed the stereochemistry at C-15 but was ambiguous for C-20 because the isolated ajmalicione, under

CHART 5.7

SOME PROPERTIES OF RING E OXYGEN HETEROCYCLES

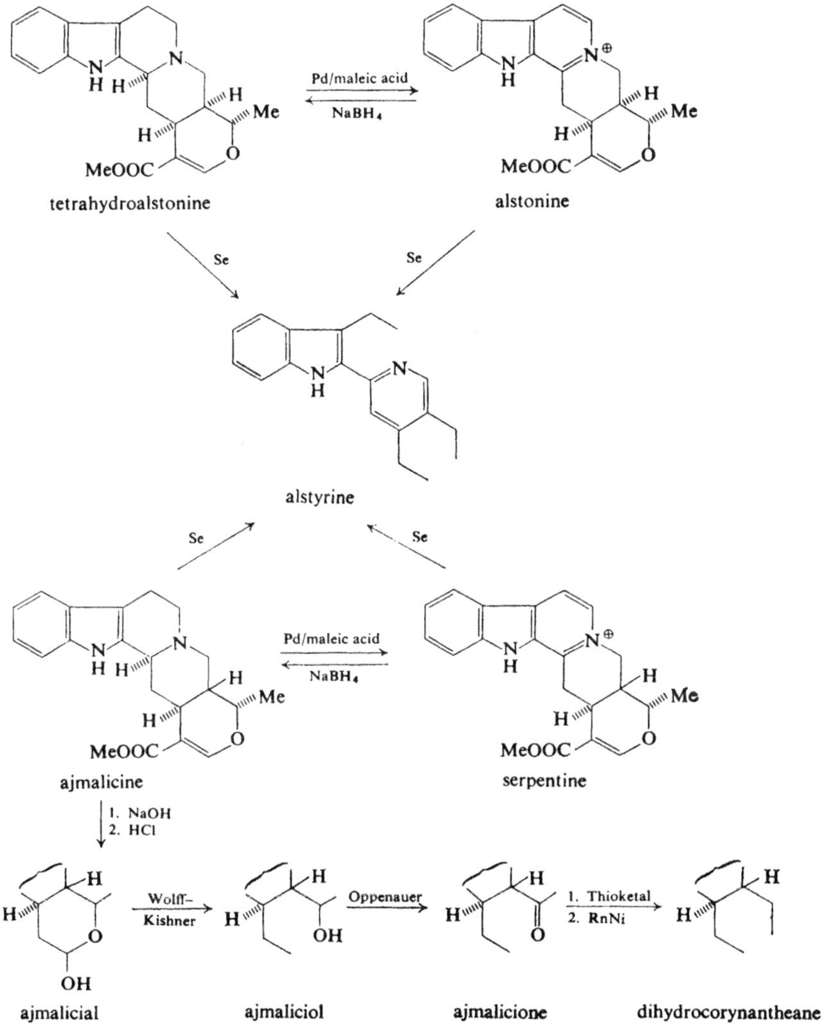

the Oppenauer conditions, could have been the epimerization product of the less stable *cis*-ketone. The stereochemistry of the DE ring junction was considered to have been established by comparing the rates of catalytic dehydrogenation (palladium, maleic acid as hydrogen acceptor) of various ring E oxygen heterocycles with the results of parallel studies carried out on the isomeric yohimbanes. These results, if all the factors were properly understood, was in clear agreement with the conclusion that ajmalicine and tetrahydroalstonine had *cis* and *trans* DE ring junctions respectively. The stereospecific total synthesis of ajmalicine reversed this conclusion, as also have measurements of rates of alkylation and the proton topology deduced from nuclear magnetic resonance spectra. The latter two physical methods are also in agreement in assigning α-configuration to the methyl on C-19 to both alkaloids. Bases are also known in which the methyl is β. It should also be pointed out here that the $pK_a = 5{\cdot}7$ of ajmalicine is about 2 units less than for yohimbine and the reason for this is not at all clear.

Oxindole alkaloids. Oxindoles equivalent to many of the *seco* and ring E oxygen heterocycles are of natural occurrence. The establishment of these relationships had been deduced entirely by the analogous chemical behaviour of the respective ring E groups but was not established rigorously until a method for converting tetrahydro-β-carboline alkaloids to their oxindole equivalents was discovered. The reverse of this conversion is still an unsolved problem. Some aspects of the chemistry of a representative of the group, mitraphylline are shown in Chart 5.8 but the reactions of ring E are not depicted because they are the same as for ajmalicine.

It turned out that the methanolysis of the 7-chloro-7H-ajmalicine (*ter.*-butyl hypochlorite on the parent indole) gave a rearranged imido ether which afforded the oxindole on treatment with acid (it should be remembered, Chart 3.4, that reaction of the 7-chloro-7-H-derivative with inorganic acid will generate the Δ^3-compound). For the analogous *cis* DE ring alkaloids the stereochemistry of the 7-halogen substituent is unfavourable to rearrangement under methanolysis conditions, however, its 7-acetoxy equivalent is

CHART 5.8

SOME TRANSFORMATIONS OF MITRAPHYLLINE AND AJMALICINE

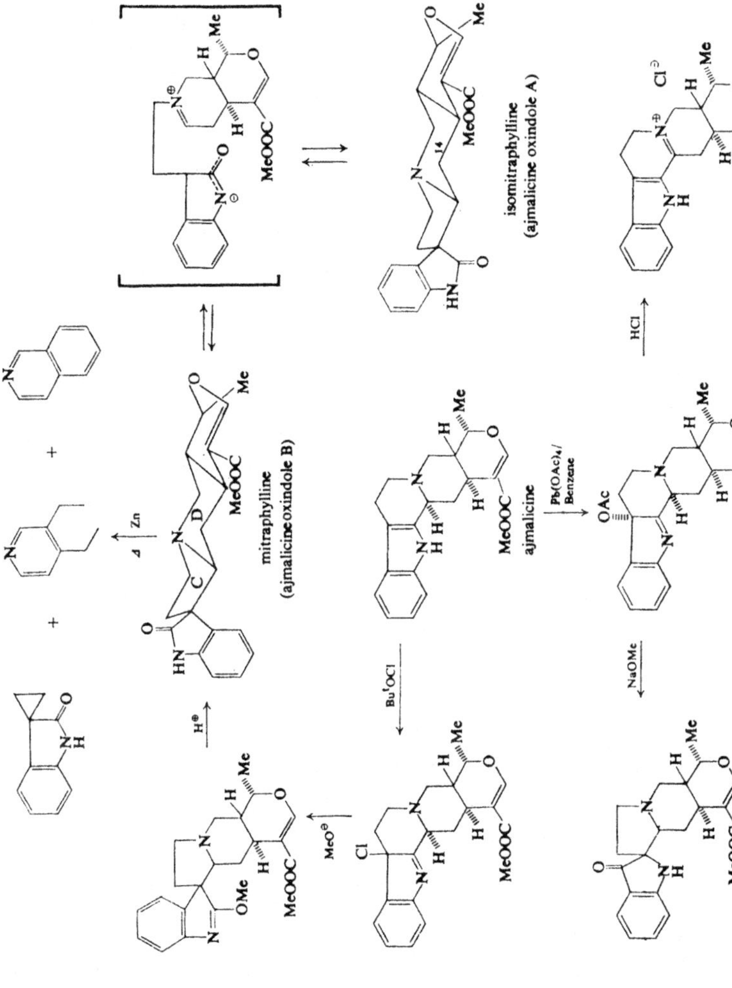

converted to the corresponding oxindole upon reflux in methanolic acetic acid. These reactions form the basis then for a chemical method for deciding the nature of the DE ring junction. It should also be pointed out here that methanolysis of 7-acetoxy compounds is a preparative method for the obtention of pseudo-indoxyls from tetrahydro-β-carboline alkaloids (Chart 5.8).

The synthesis of ajmalicine oxindole is ambiguous as far as the stereochemistries at C-3 and C-7 are concerned because the reaction conditions are such as to produce both the known readily interconvertible isomers, mitraphylline and isomitraphylline. Although four isomers are possible on theoretical grounds, apparently only the two with a *trans* CD ring fusion have been detected and isolated. Mitraphylline (ajmalicine oxindole B, the stronger base) is favoured in acidic media probably because of the extra stability produced by hydrogen bonding between the conjugate acid and the oxindole carbonyl group. In basic solution such as refluxing pyridine this stabilizing effect has disappeared and a considerable amount of isomitraphylline (ajmalicine oxindole A) is formed. These structural assignments have received support from the nuclear magnetic spectra where the C-14 methylene protons in oxindole A are subjected to shielding by the aromatic group with the resultant upfield displacement of their chemical shift relative to that of the same protons in oxindole B

SUGGESTED READING

Shamma, M. and Richey, J. M., The stereochemistry of the heteroyohimbine alkaloids, *J. Amer. Chem. Soc.*, **85**, 2507 (1963).

van Tamelen, E. E. and Hester, J. B., The total synthesis of *dl*-dihydrocorynantheine, *J. Amer. Chem. Soc.*, **81**, 3805 (1959).

van Tamelen, E. E., Shamma, M., Burghstahler, A. W., Wolinsky, J., Tamm, R. and Aldrich, P. E., The total synthesis of *dl*-pseudoyohimbine, *J. Amer. Chem. Soc.*, **80**, 5006 (1958).

Witkop, B., Zur Konstitution des Yohimbins und seiner Abbauprodukte, *Justus Liebigs Ann. Chem.*, **554**, 83 (1943).

Woodward, R. B., Bader, F. F., Bickel, H., Frey, A. J. and Kierstead, R. W., The total synthesis of reserpine, *Tetrahedron*, **2**, 1 (1958).

CHAPTER VI

THE STRYCHNOS ALKALOIDS

THE principal members of this group (Chart 6.1) are derived mainly from the Wieland-Gumlich aldehyde (or its equivalent). An alternate ring closure gives the closely related aspidospermatine class of which condylocarpine is a member. None of the biosynthetic steps are known but might proceed according to the experimentally realized *bis*-Mannich condensation (Chapter III, p. 25) equivalent to the hypothetical scheme in Chart 2.2 or by the routes sketched in Charts 2.3 or 12.2.

The alkaloids of this group were first discovered in the genus, *Strychnos* (family, *Loganiaceae*) but strangely the dimeric curare alkaloids (tertiary and quaternary) occur only in the South American species. In the *Apocynaceae*, e.g. picralima and vinca species, products of the further oxidation of the Wieland-Gumlich aldehyde are found. The Wieland-Gumlich aldehyde shows a great tendency to dimerize which may in the absence of other enzymatic pathways in a particular plant make it almost mandatory for it to be isolated as a dimeric derivative. Such a doubling up can be prevented by acetylation (diaboline) oxidation (akuammicine) or reduction (spermostrychnine).

The chemistry of strychnine has not played as direct a role in the development of heterocyclic chemistry as morphine because its structure for so long defied exact description. It did, however, generate some important theoretical questions and was there, for half the twentieth century as one of the great chemical challenges and upon it almost every useful reaction in degradative chemistry has been applied usually with some success. Although its structure is now known with certainty by degradation, synthesis and X-ray

Chart 6.1
Representative strychnos alkaloids

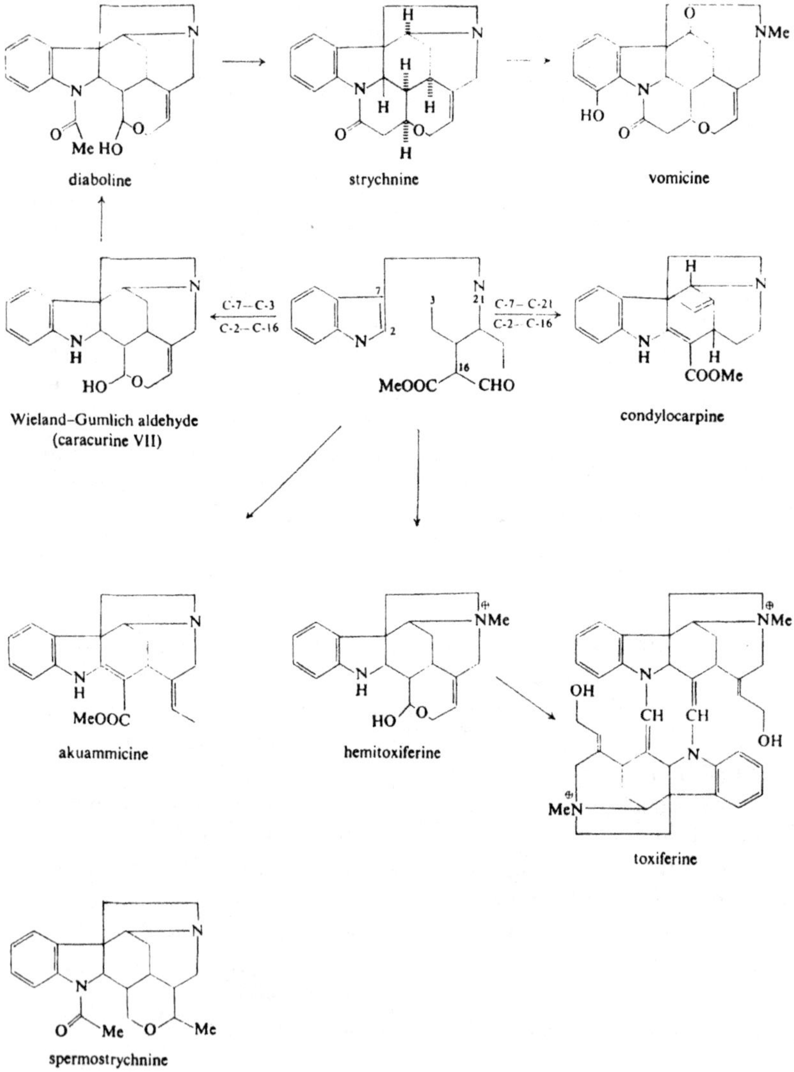

crystallography many of its reactions deserve renewed study for their mechanistic implications. Practically all the compounds of this group are interconvertible and since strychnine is commercially available and cheap it is the conventional starting point for model studies or partial syntheses.

STRYCHNINE

The carbon patterns of the products (Chart 3.1) of the drastic degradation of strychnine can all be discerned in the parent molecule but cannot by themselves be used to deduce a unique formula. That the A and B rings were six and five membered respectively was reconfirmed over the years often at the cost of considerable labour. One such case concerned dinitrostrycholcarboxylic acid, one of the nitric acid oxidation products of strychnine. It was first obtained about the turn of the century and after considerable work in the late twenties was found in the early thirties to be 5,7-dinitroindole-2-carboxylic acid. Actually the structure of strychnine would probably have been realized much earlier than it was if any one of a number of degradations had been persevered with in a systematic way. The constitution arrived at by the chemical methods rests on the properties of the functionalities in their special environments and their interlocking reactions. The advent of commercial recording infrared and ultraviolet machines played an important part in the latter phase of this work. A synthesis of the alcohol, isostrychnine I (strychnine with its cyclic ether opened at dotted line and $\Delta^{16,17}$ double bond) has confirmed these conclusions as has the determination of the structure and absolute stereochemistry by the X-ray crystallographic method.

Much of the chemistry of strychnine is a consequence of two factors, the environment of the bridgehead basic nitrogen which makes it different from a simple tertiary amine and the possibility of base induced reactions made possible by the lactam carbonyl in ring G. For a proper understanding of strychnine chemistry its topology must be remembered. From a theoretical point of view there are many possibilities stemming from the ability to form the enolate anion at C-22 and some of these are indicated in Chart 6.2. As expected none of them occur once the carbonyl is reduced to furnish strychnidine. Where —OR is initially present as a cyclic ether the product(s) were, according to the conditions, found to be hydroxy unsaturated lactams with, or without, stereochemical change at C-16 and/or the aminoacid with or without changes in

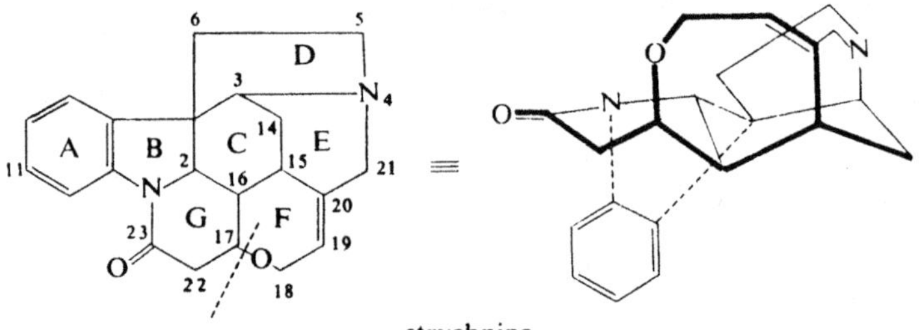

strychnine

stereochemistry at C-16 and/or C-17. The products of the acidic ether fission of strychnine can be related by hydrolysis to the corresponding base catalysed degradation products. It should be noted that the generation of the N-phenylpyridone derivatives (readily diagnosed by their characteristic ultraviolet absorption spectra) indicated that protons must be attached to C-2 and C-16. This deduction was made certain by catalytic reduction of a suitable N-phenylpyridone back to a known tetrahydro-compound.

The second order Beckmann rearrangement carried out on the C-22 oxime of strychnine is the first step of a preferred pathway for the conversion of the alkaloid into the Wieland-Gumlich aldehyde (Chart 6.2).

The above properties of ring G were of value in establishing the chain of atoms between the two nitrogens after oxidative fission of the double bond (Chart 6.3). A complicating feature of the suffix "a" compounds was their propensity to furnish the two other theoretically possible double bond isomers in basic media.

CHART 6.2

IMPORTANT PROPERTIES OF RING G OF STRYCHNINE AND ITS DERIVATIVES (PARTIAL FORMULAE)

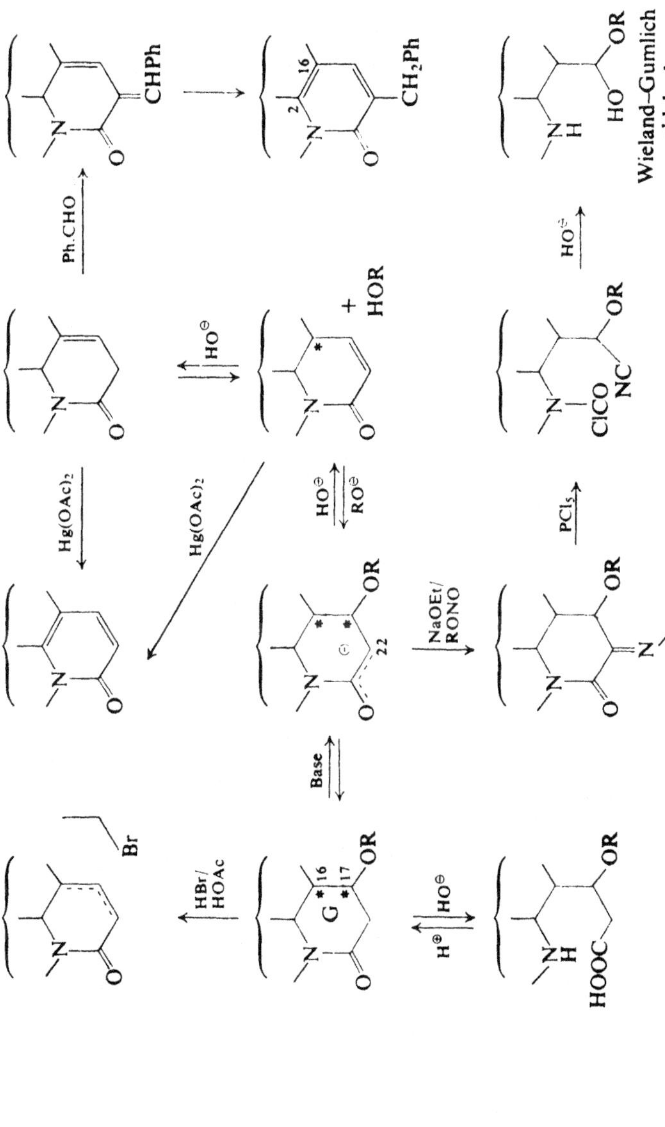

CHART 6.3
OXIDATIVE DEGRADATION OF STRYCHNINE

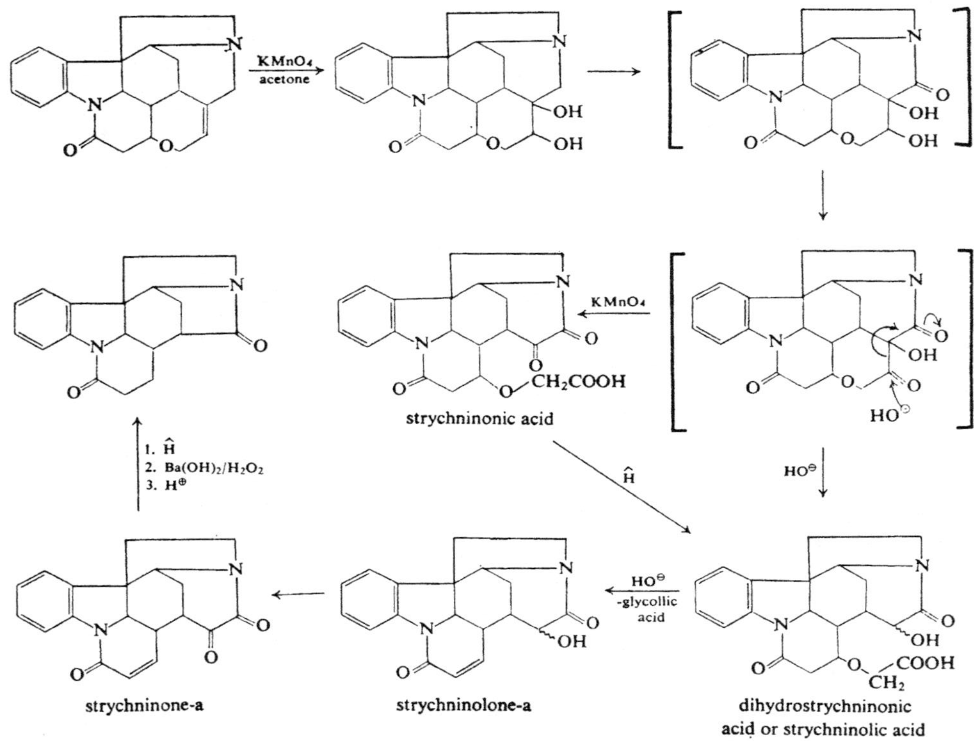

Neostrychnine. Neither the Hofmann, Emde, or cyanogen bromide reactions have been particularly useful as starting points for systematic degradations of strychnine because of the stereochemistry which is not suitable for eliminations and instead a multiplicity of S_N2 products were usually formed. It was a study of the Hofmann degradation that first led to neostrychnine (Chart 6.4), a double bond isomer ($\Delta^{20,21}$) of strychnine, a compound more readily made by heating strychnine with selenium or Raney

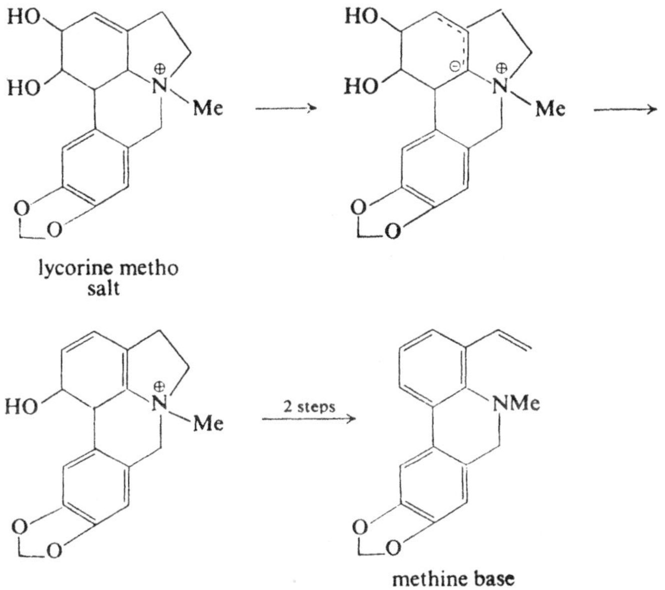

nickel. Thus strychnine methosulfate upon reflux in methanol containing an alkali hydroxide afforded a mixture of strychnine and N-methyl-5-methoxy-4,5-dihydroneostrychnine via double bond rearrangement made possible through the intermediacy of the ylide (hydrogen abstraction at C-21) followed by an S_N2 displacement at C-5 by methoxide ion (an analogous "ylide" is also an intermediate in the first step of the Hofmann degradation of lycorine). The N-methylmethoxydihydroneostrychnine upon reflux in dilute acid afforded a neostrychnine metho salt, pyrolysis of which gave neostrychnine itself. The low basicity of neostrychnine (pK_a 3·72) has been attributed to an inductive effect of the double bond which cannot undergo a prototropic shift to form an iminium salt without violating Bredts' rule. An alternative reason may well be that the rigidity of the system makes it difficult for the nitrogen to assume a strain-free tetrahedral configuration required for its conjugate acid.

CHART 6.4
RELATIONSHIP BETWEEN STRYCHNINE AND NEOSTRYCHNINE

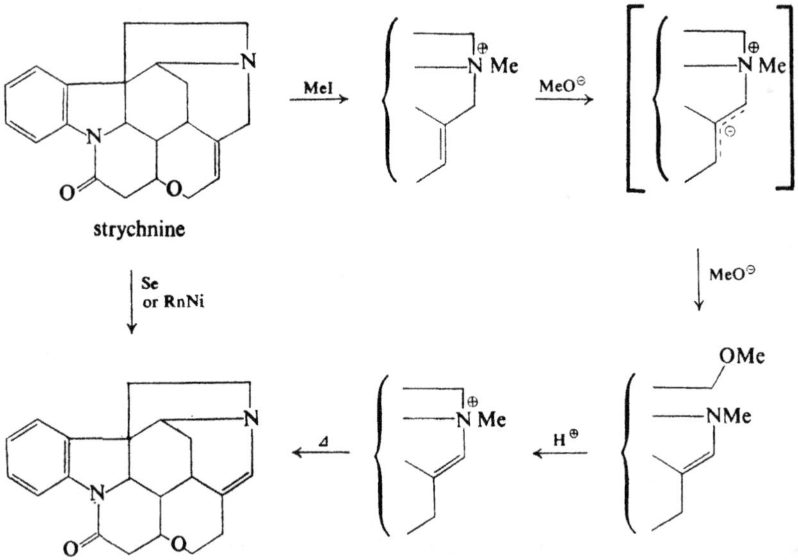

neostrychnine

N-methyl-5-methoxy-4,5-dihydroneostrychnine

Both neostrychnine and N-methyl-5-methoxy-4,5-dihydroneostrychnine were oxidized by potassium permanganate with insertion of two oxygen atoms. One oxygen was clearly a keto function and the other apparently a carbonyl in a lactam rather than an amide because it was stable to reflux in strong acid. The situation was further complicated by the discovery that perbenzoic acid

CHART 6.5

SOME PROPERTIES OF NEOSTRYCHNINE (PARTIAL FORMULAE)

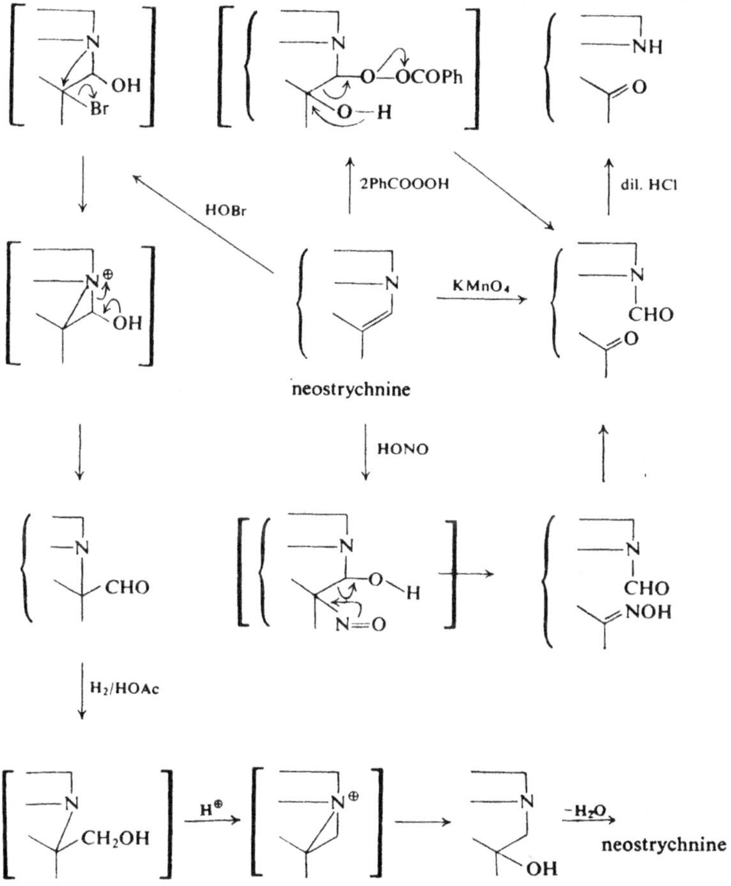

(2 moles) gave the same product and that nitrous acid and diazotized *p*-nitroaniline afforded the corresponding oxime and

p-nitrophenylhydrazone respectively. The mystery was largely resolved when by switching to dilute hydrochloric acid the oxidation product furnished formic acid (Chart 6.5). With bromine water neostrychnine underwent yet another odd reaction to produce an aldehyde. This aldehyde gave a stable alcohol upon reduction in neutral medium but in acetic acid a labile alcohol was produced which readily lost water to regenerate neostrychnine. Possible courses for these reactions of neostrychnine are given in Chart 6.5 and it cannot be emphasized too strongly that they are determined by the stereochemistry in which the nitrogen and the double bond have found themselves.

Pseudostrychnine. This compound was first isolated from the mother liquors of strychnine manufacture but it is probably an artefact arising from the aerial oxidation of the base. A good preparative method is to heat strychnine N-oxide in the presence of a little potassium chromate. Although it is a carbinolamine and interconvertible into a non-isolable keto-amine (*sec*-pseudostrychnine) it apparently does not afford isolatable iminium salts. It does however form ethers simply by crystallizing pseudostrychnine from the chosen alcohol and they are as readily decomposed in dilute mineral acid. The N-methyl derivative, N-methyl*sec*pseudostrychnine, was an important model compound in the study of vomicine, in fact the close relationship was first unequivocally established by the isolation of the same product after the oxidative removal of rings A and B. N-Methyl*sec*-pseudostrychnine shows properties typical of a methylene ketone e.g. yielding oxime and benzylidene derivatives and because of the stereochemistry undergoing transannular reactions with the nitrogen—see Chart 6.6. A most interesting property of pseudostrychnine was its oxidation to strychnone, an N-acylindole, which defined the third linkage between the two nitrogens of strychnine, the other two being provided by the isolation of tryptamine (Chart 3.1) and the results of oxidative degradation (Chart 6.3). Some of these properties of pseudostrychnine are illustrated in Chart 6.6.

CHART 6.6

SOME PROPERTIES OF PSEUDOSTRYCHNINE (PARTIAL FORMULAE)

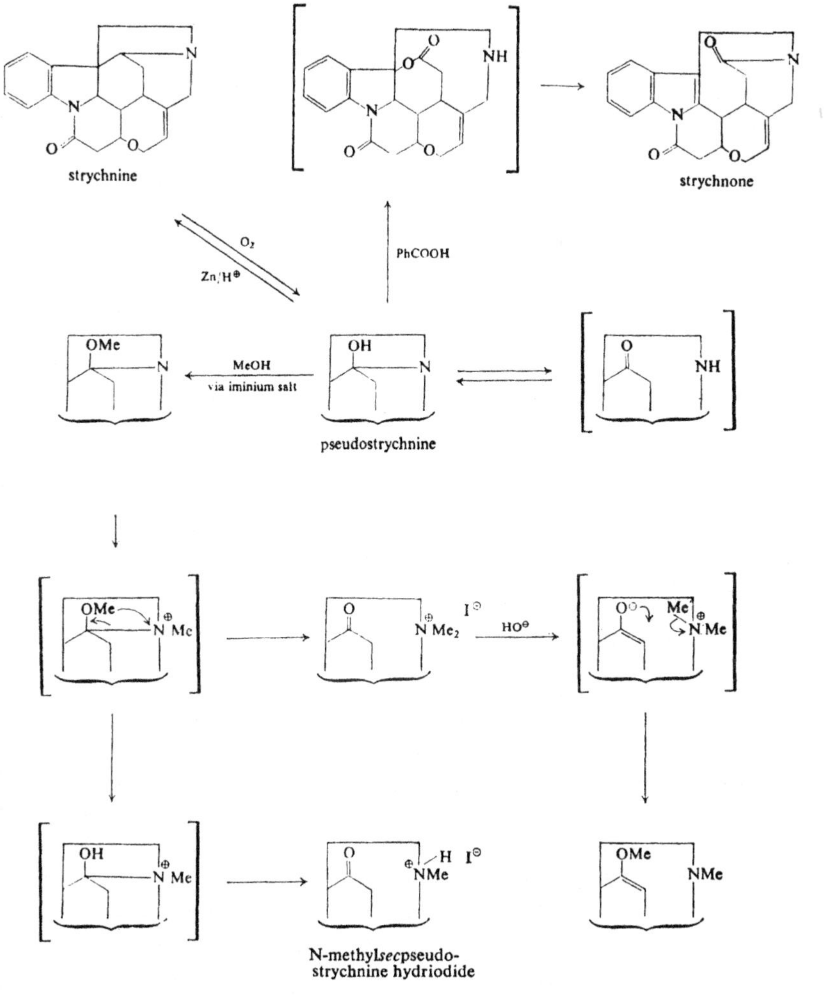

AKUAMMICINE

The structure of this alkaloid was first put forward on the basis of its ultraviolet absorption spectrum being that of a β-anilino-acrylate and the products of drastic degradation. This conclusion was confirmed not only by other degradations but also by chemical interconversions with the Wieland–Gumlich aldehyde. Akuammicine is remarkable for the ease with which it breaks up into aromatic optically inactive components (Chart 6.7). Upon heating in a sealed tube in methanol at 80° the "betaine" is formed and at 140° in the same solvent it furnished 3-ethylpyridine and 2-hydroxy-carbazole in very high yield. A mechanism for this process was already given in Chart 3.2. Racemic akuammicine (pseudoakuammicine) has also been obtained from the same plant, *Picralima klaineana*, and could conceivably have been an artefact of the isolation arising by reclosure of one of the early symmetrical intermediates in the decomposition described above. This may be so but akuammicine has never been successfully racemized in the laboratory so that pseudoakuammicine must be for the present regarded as an exception to the observation that all Type I alkaloids have the same stereochemistry at C-15.

Upon acid hydrolysis akuammicine like a β-ketoester suffered decarboxylation. The resulting indolenine exists in equilibrium with the indole iminium salt (a reverse of a Mannich condensation, cf. the synthesis of gramine, Chart 4.1; see also epimerization of C-3 in the tetrahydro-β-carbolines, p. 26). In the absence of a hydrogen donor the indolenine form is favoured *e.g.* reduction with lithium aluminium hydride in ether gave the indoline (dihydro-indole) whereas in methanol (hydrogen donor) reduction with potassium borohydride afforded the indole. This sort of chemistry has since been recognized in other similarly constituted indole alkaloids and has been used not only for diagnostic purposes but also to realize interconversions and syntheses (cf. iboga and aspidosperma alkaloids) and may be important in biosynthesis. Stemmadenine is an example of a naturally occurring ring opened species (Chart 6.7).

CHART 6.7
SOME PROPERTIES OF AKUAMMICINE

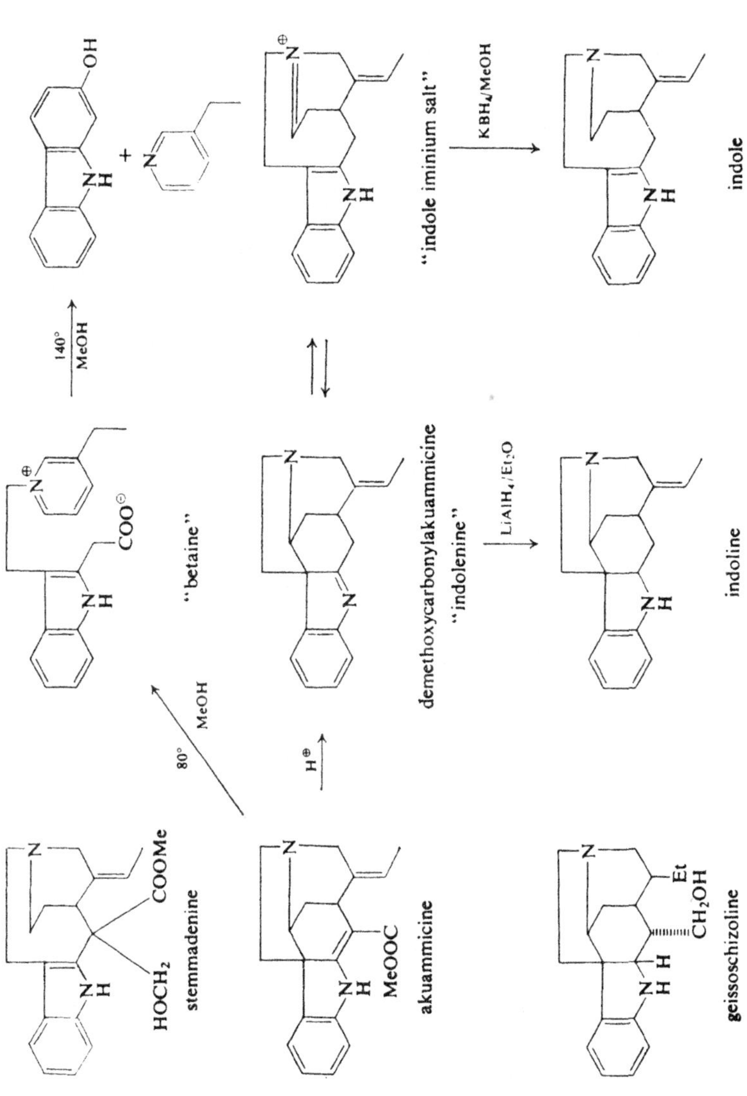

SUGGESTED READING

BERNAUER, K., Alkaloide aus Calebassencurare und Südamerikanischen Strychosarten, *Fortschritte der Chemie Organischer Naturstoffe*, **17**, 183 (1959).

LEUCHS, H., Zur Kenntnis der Strychnos-Alkaloide. 1 Mitteilung. Oxydation des Brucins und Strychnins nach einer neuen Methode, *Ber. dtsch. chem. Ges.*, **41**, 1711 (1908).

NAGARAJAN, K., WEISSMANN, CH., SCHMID, H. and KARRER, P., Synthese der racemischen und optisch aktiven Formen des 16α-Strychindols; chemische Ableitung der absoluten Konfiguration des Strychnins, *Helv. chim. Acta*, **46**, 1212 (1963).

PEERDEMAN, A. F., Absolute configuration of natural strychnine, *Acta Cryst.*, **9**, 824 (1956).

SMITH, G. F. and WROBEL, J. T., Akuammicine, *J. Chem. Soc.*, **1960**, 792.

WIELAND, H. and GUMLICH, W., Ueber einige neue Reaktionen der Strychnos Alkaloide, *Justus Liebigs Ann. Chem.*, **494**, 191 (1932).

WOODWARD, R. B., CAVA, M. P., OLLIS, W. D., HUNGER, A., DAENIKER, H. U. and SCHENKER, K., The total synthesis of strychnine, *Tetrahedron*, **19**, 247 (1963).

CHAPTER VII

CINCHONAMINE AND QUINAMINE

THERE are recognized at present three naturally occurring members of this group, cinchonamine, quinamine and conquinamine (3-epiquinamine), minor alkaloids of certain cinchona species. Elucidation of their structures lead to the suggestion that the cinchona bases may be derived biosynthetically from Type I precursor and tryptophan. It has since been shown that isotopically labelled tryptophan is incorporated into these quinoline bases. It is convenient to point out here some other alkaloids in which Type I precursor may be dissected out, e.g. the *bis*isoquinoline base, emetine; the indoleisoquinoline, tubulosine; and also the indoles uleine and ellipticine which appear to have lost (or never had) the tryptophan side chain. (Chart 7.1)

Cinchonamine. Early work had shown that cinchonamine gave colour reactions indicative of an indolic nucleus in contrast to the co-occurring cinchona bases and it differed in another way by yielding 3-vinylquinuclidine-8-carboxylic acid upon chromic acid oxidation. The selection of an efficient route for getting at the remaining portion of the molecule was greatly facilitated by the use of an ultraviolet spectrophotometer as a quantitative detector of changes in the chromophoric moiety during and after an experiment. When it was discovered that the vigorous acetylation product of cinchonamine gave an ultraviolet absorption spectrum indicative of a vinylindole enough of the derivative was prepared for a permanganate oxidation which gave 3-β-acetoxyethylindole-2-aldehyde (Chart 7.2).

Physical methods were also of great help in clarifying the nature

CHART 7.1

FURTHER EXAMPLES OF ALKALOIDS CONTAINING THE TYPE I PRECURSOR

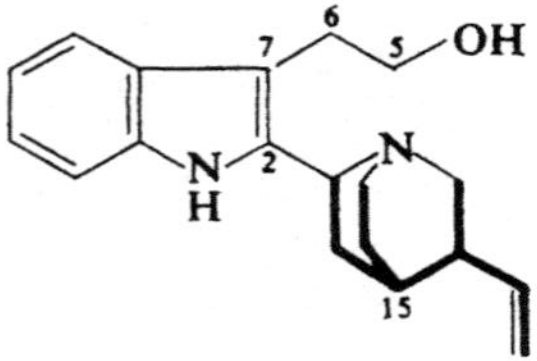

cinchonamine

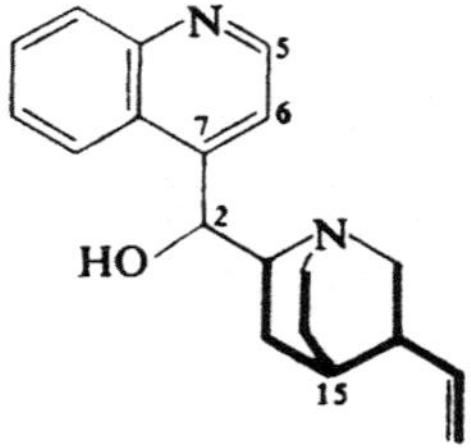

cinchonine

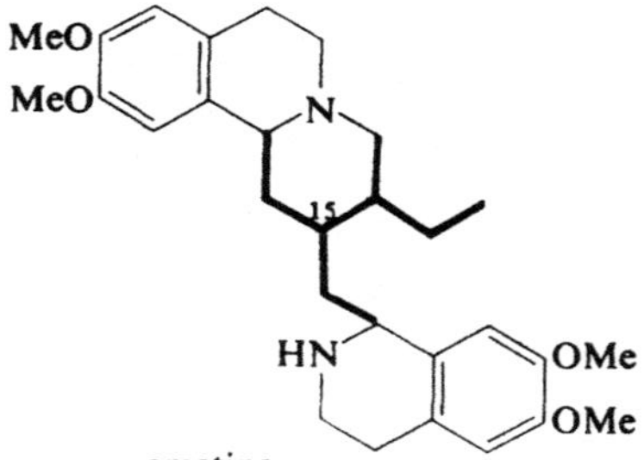

emetine

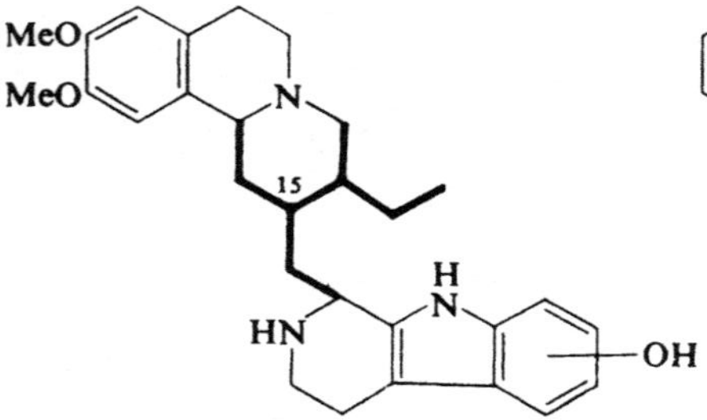

tubulosine

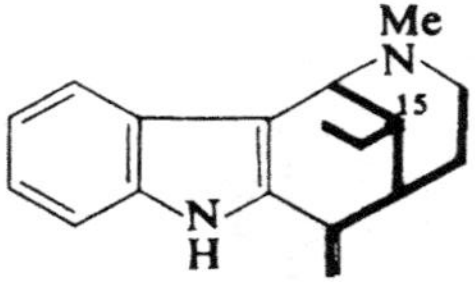

uleine

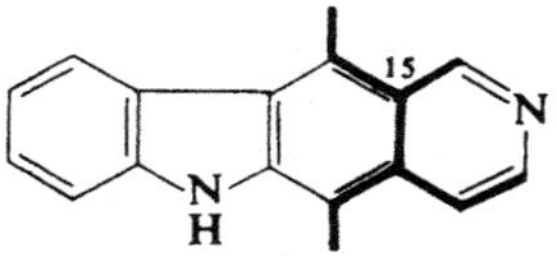

ellipticine

CHART 7.2
SOME PROPERTIES OF CINCHONAMINE

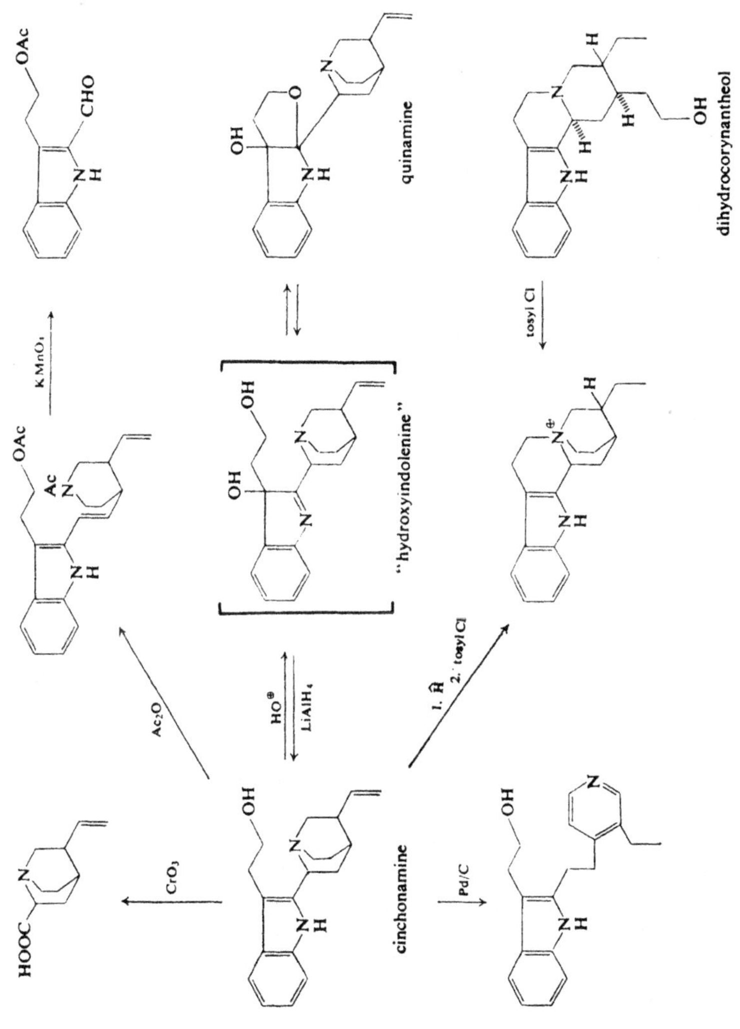

of the palladium catalysed dehydrogenation of cinchonamine. The product analysed for two protons less than the starting material but deep seated changes must have occurred since the product was a weak base (pK_a titration) and had an ultraviolet absorption spectrum consistent for nonconjugated indole and pyridine chromophores.

The absolute stereochemistry (with the exception of C-3) of dihydrocinchonamine has been secured by its synthesis from dihydrocinchonine (reversal of the latter's formal biogenesis!) Dihydrocinchonamine and dihydrocorynantheol have been converted into the same quaternary derivative which settled the C-3 stereochemistry if no epimerization (see p. 26) had taken place under the experimental conditions.

Quinamine. The structure of quinamine follows from its reduction with lithium aluminium hydride to cinchonamine and resynthesis by treating the latter alkaloid with perbenzoic acid. The properties of quinamine under a variety of conditions are those implied in Chart 3.5, the intermediate here being the hydroxyindolenine. It is therefore easy to understand why treatment of quinamine with base under a variety of conditions gave a mixture of products. Isomerization at C-3, i.e. formation of 3-epiquinamine, is due to the equilibrium "indolenine"$\rightleftarrows$"methyleneindoline". Upon prolonged action of base, the pseudoindoxyls (the isoquinamines) are the logical end products but their stereochemistries at C-3 cannot be unequivocally deduced until it is known which of the equilibrating quinamine isomers rearranges the fastest.

When quinamine was refluxed in dilute acetic acid for a prolonged period it afforded the ketoamine, quinamicine, which may have been the result of a direct loss of water or more likely an allylic rearrangement of the tertiary hydroxyl group in the "hydroxymethyleneindoline" form then elimination (the hydroxyl cannot be lost to form the iminium ion because of the bridgehead position of the quinuclidine nitrogen). The formulation of quinamicine is the only reaction of these compounds which has any parallel in

CHART 7.3
TRANSFORMATIONS OF QUINAMINE

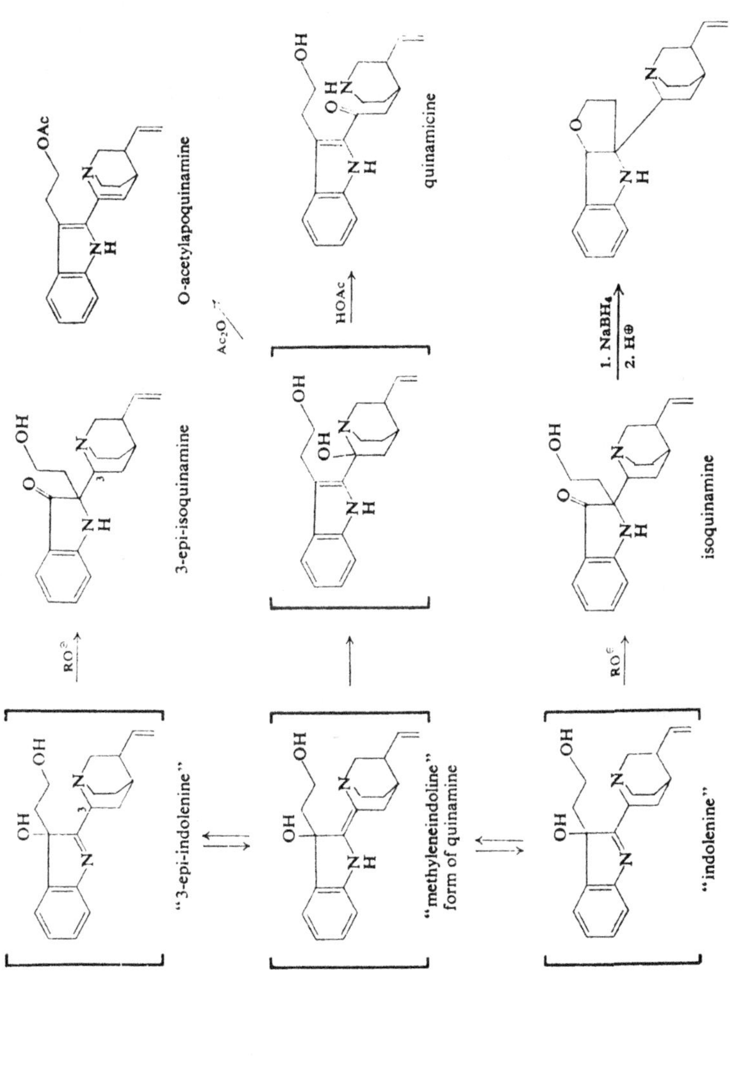

cinchona alkaloid chemistry. It is also not difficult to rationalize how acetic anhydride must generate O-acetylapoquinamine from quinamine (Chart 7.3).

Pseudoindoxyls after reduction usually rearrange in acidic solution back to indoles, but it is easy to see for the particular case of dihydroisoquinamine that cyclization to a tetrahydrofuran would be the preferred pathway. The problem then is changed to what acidic conditions are required to split open this particular cyclic benzyl ether.

In the course of the early work it was observed that quinamine heated above its melting point evolved formaldehyde, at the time this was taken to mean that the alkaloid had a hydroxymethyl on the indole α-carbon. In the light of the accepted formula it is more likely that this aldehyde is a special case of a pyrolysis of a 1,3-glycol that is, a retro-Prins reaction.

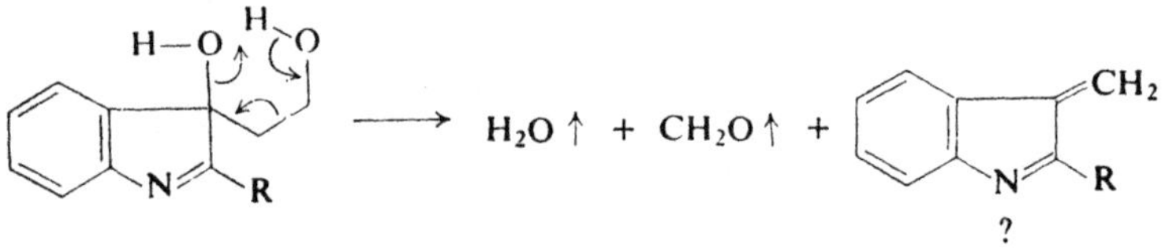

SUGGESTED READING

Ochiai, E., Ishikawa, M. and Oka, Y., The formation of dihydrocinchonamine from cinchonidine, *Chem. pharm. Bull.*, 7, 744 (1959).

Witkop, B., Réarrangements de cétones, alcamines, peroxydes, ozonides et alcaloïdes dans la série de l'indole, *Bull. soc. chim. Fr.*, **1954**, 423.

CHAPTER VIII

THE AJMALINE–SARPAGINE BASES

DETAILED knowledge of the ajmaline–sarpagine group of alkaloids is of quite recent origin, being intimately connected with the intensive investigation of rauwolfia bases brought about by the commercial usefulness of reserpine. As the structure of ajmaline began to unfold it was guessed to be a new variant of Type I precursor and tryptophan and since this assumption was correct it helped to shorten the time required to arrive at the correct formula by severely restricting the number of possibilities to be considered. The alkaloids of this group have been interrelated and they form a self-consistent set whose structures rest on three separately derived proofs. Two of these were X-ray crystallographic analyses of sarpagine derivatives and the third was a facile unfolding of the ajmaline skeleton to ring E *seco* derivatives of the complex tetrahydro-β-carboline alkaloids. It should be noted that like strychnine much of the relative stereochemistry is simply a consequence of the existence of the molecule. Thus of the ten asymmetric centres in ajmaline the six in the nucleus are determined automatically which leaves only the four peripheral ones to worry about. It has transpired that the ajmaline congeners isolated from non-rauwolfia plants, e.g. *Vinca* have a 2-epi configuration.

Ajmaline. The presence of the indoline was established by colour reactions, substitution reactions, e.g. bromoajmaline, and by ultraviolet absorption spectroscopy. Its N-methyl was fixed onto the indoline nitrogen since potassium permanganate oxidation in *acetone* gave N-methylisatinacetone and soda lime or zinc dust distillation furnished N-methylharman.

It was during the study of the second nitrogen of ajmaline that its interesting properties emerged (Chart 8.1). It turned out to be a carbinolamine with a bridgehead nitrogen (cf. pseudostrychnine

CHART 8.1

SOME REACTIONS OF AJMALINE ABOUT N-4 (PARTIAL FORMULAE)

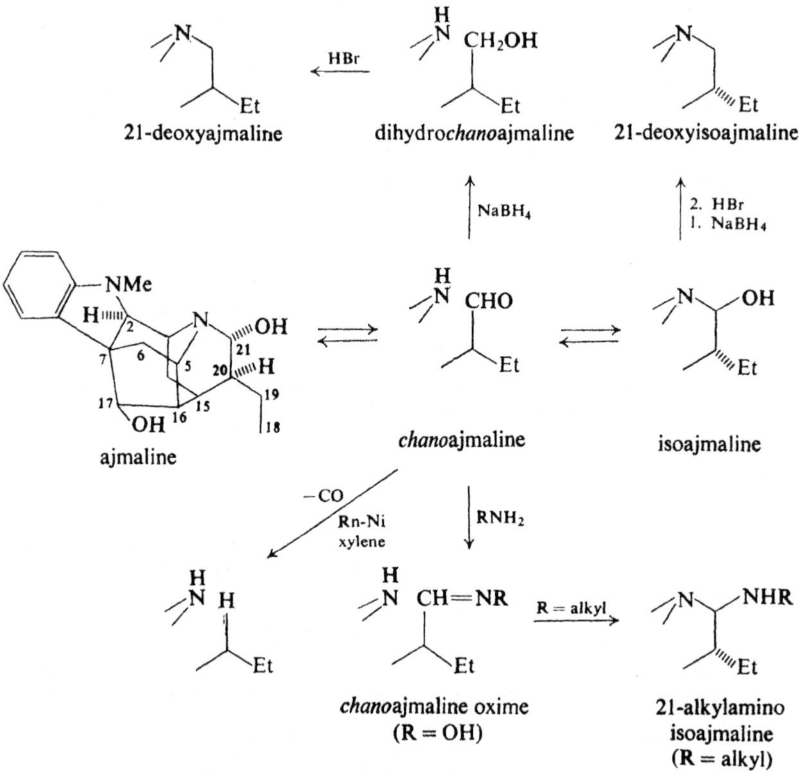

and some of the properties of quinamine) hence it could not be converted into an iminium ion and therefore had many of the normal reactions of a tertiary amine and an unhindered secondary alcoholic function. However, it also exhibited other properties better interpreted as taking place via the tautomeric amino aldehyde which will be referred to as *chano*ajmaline. Thus ajmaline although showing no infrared band indicative of a carbonyl function did form an oxime which could be dehydrated to yield a nitrile. It also underwent a Wolff–Kishner reduction to deoxydihydro*chano*ajmaline (—CHO→—CH_3) and was reduced by sodium borohydride to dihydro*chano*ajmaline (—CHO→—CH_2OH), whose mineral acid salts upon heating lost water to yield 21-deoxyajmaline. The presence of this masked carbonyl function very neatly accounts for isoajmaline (epimeric ethyl group) formed when ajmaline is refluxed in a basic medium or heated above its melting point.

In an unusual reaction ajmaline refluxed with Raney-nickel in xylene decarbonylates and this observation has been of use in the isolation of the C-21 atom in biosynthetic work.

For the study of the remainder of the molecule 21-deoxyajmaline was the most useful compound. Except for the stereochemistry at C-2, -17, -20 and -21 the absolute stereochemistry of the ring system was defined as a result of the chemical unfolding (Chart 8.2) of 21-deoxyajmaline to 1-methyltetradehydrocorynantheidane and 21-deoxyisoajmaline to 1-methyltetradehydrodihydrocorynantheane (corynantheane had previously been correlated with cinchonine, Chapter VII). The first step was best accomplished by oxidation with lead tetraacetate in a non-polar medium probably to an indolenine which opens up in a manner analogous to demethoxycarbonylakuammicine going to the indole (Chart 6.7). Because of diaxial interactions the A-aldehyde was readily epimerized to the less hindered B-aldehyde. Reduction of this compound, tosylation followed by an elimination, and disproportionation gave the tetradehydro compound. Whether this elimination proceeded via the participation of the electron pair on N-4 or loss of the proton at C-6 is not known.

That the 17-hydroxyl group was secondary and in a five membered ring was shown by its Oppenauer oxidation to a ketone (infrared spectrum indicative of a cyclopentanone). Reduction of this ketone would have been expected to proceed by addition of the hydrogen on the unhindered face that is the one before the

CHART 8.2

DEGRADATION OF 21-DEOXYAJMALINE

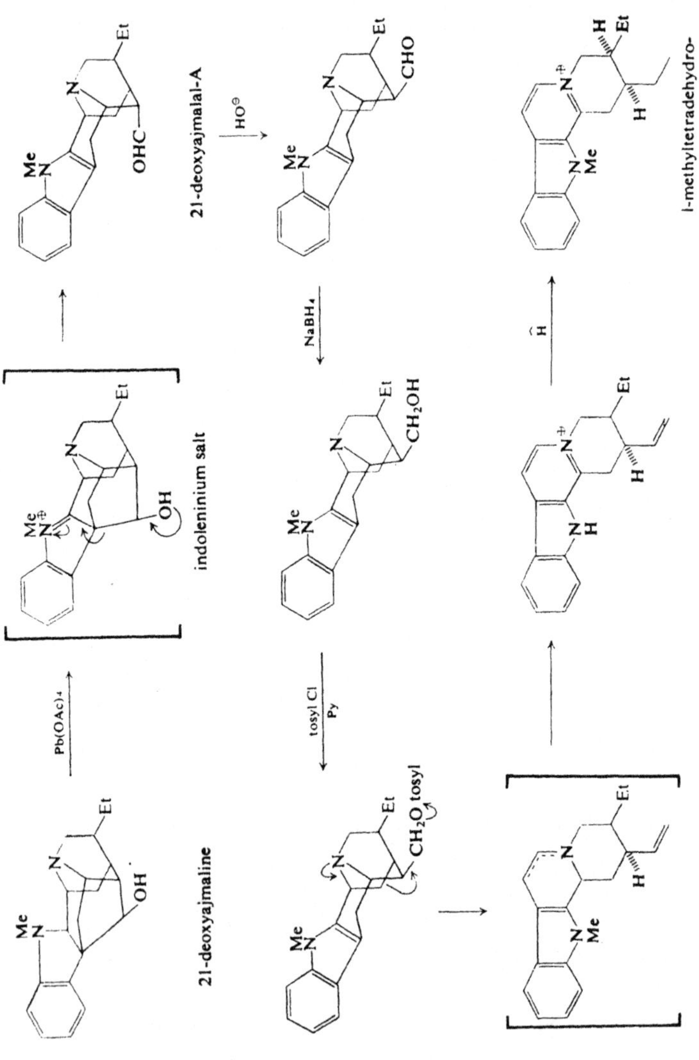

viewer (in the projection drawings). Since an epimeric alcohol was produced, the stereochemistry of 17-hydroxy group in 21-deoxyajmaline is correctly pictured. A second proof depended on the measurement of the coupling constant (J) for the interaction between the C-17 and C-16 protons in the epimeric alcohols. For the original alcohol (J) was less than 2 cps, i.e. the angle between the two protons is about 90° and for the 17-epimer $J = 9$ cps and both results were in agreement with prediction.

The stereochemistry at C-2 followed from the preparation of the other epimer (Chart 8.3). This was accomplished by the oxidation of 21-deoxyajmaline acetate with an excess of lead tetraacetate which gave an indolenine whose reduction and N-methylation gave 2-epi-21-deoxyajmaline-17-O-acetate. Since catalytic hydrogenation would have been expected as in the case of the 17-ketone to have proceeded from the topside of the molecule it would indicate that in the original, N-1 was axially oriented on ring C. Comparison of the dissociation constants of the epimeric pairs was in agreement with this conclusion, 21-deoxyajmaline with the possibility for the proton of the conjugate acid to hydrogen bond to N-1 had pK_a 8·44 which was 0·64 units higher than for its C-2 epimer where such a bond was not possible. As expected the optical rotatory dispersion curves of the two compounds were virtually in a mirror image relationship to each other.

Because of lessened steric interactions the configuration of C-21 hydroxyl group would be expected to be *trans* to the neighbouring C-20 ethyl since equilibration is possible via the aldehyde of the *chano* species. It is not at all obvious why ajmaline should isomerize so completely to isoajmaline, although examination of accurate models would suggest that the quinuclidine moiety is skewed with the result that the ethyl of ajmaline should be more hindered than its epimer. The nuclear magnetic resonance spectral results supported the above stereochemical assignments for ajmaline but were ambiguous for isoajmaline where the 20 and 21 protons have about the same theoretical coupling constant no matter which configuration of the 21-hydroxyl is chosen.

CHART 8.3

CONVERSION OF 21-DEOXYAJMALINE INTO THE 2-EPI-COMPOUND

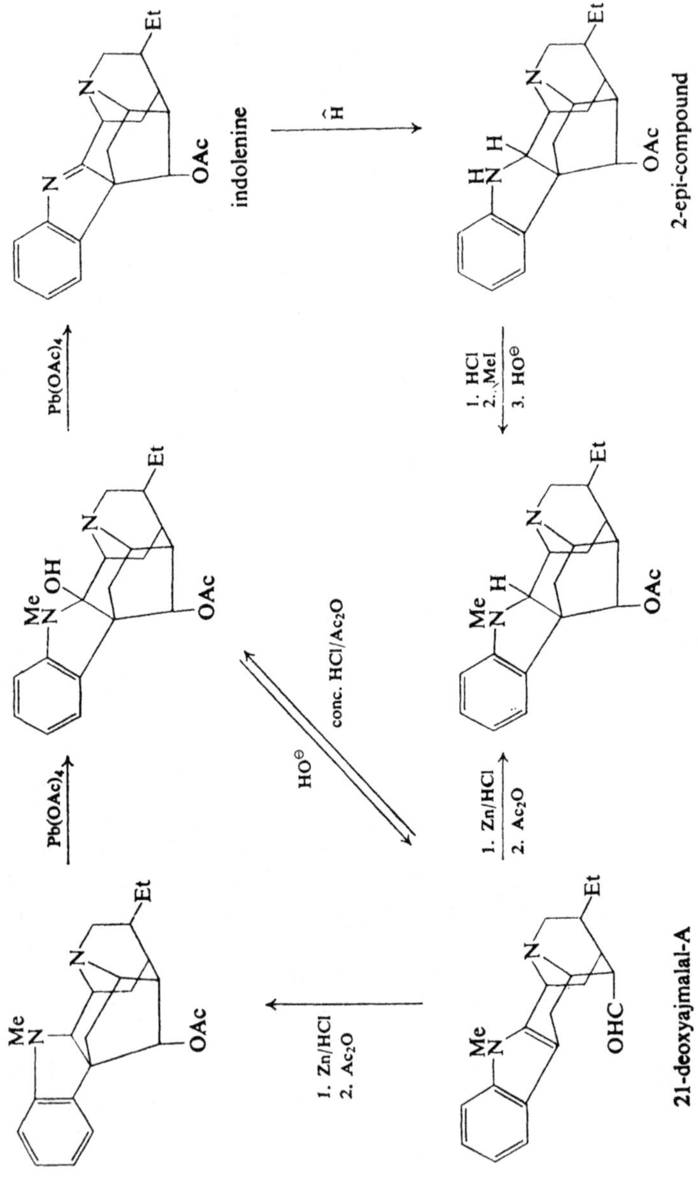

From a synthetic and biosynthetic point of view it is of interest to note that 21-deoxyajmalal-A in strongly acidic solution exists in the ring closed indolenium form (Chart 8.3; cf. demethoxycarbonylakuammicine, Chart 6.7) which can be trapped either by acetylation or reduction (zinc). Although a trace of 21-deoxyajmaline was formed in these experiments none of the more hindered 17-epi-compounds were detected.

The ease with which advantage can be taken of the nucleophilic reactivity of the indolic β-position toward a suitably placed group was further illustrated by the behaviour of 17-O-tosyl-21-deoxyajmalol-A which even upon heating gave 2-hydroxy-2-epi-17,21-dideoxyajmaline.

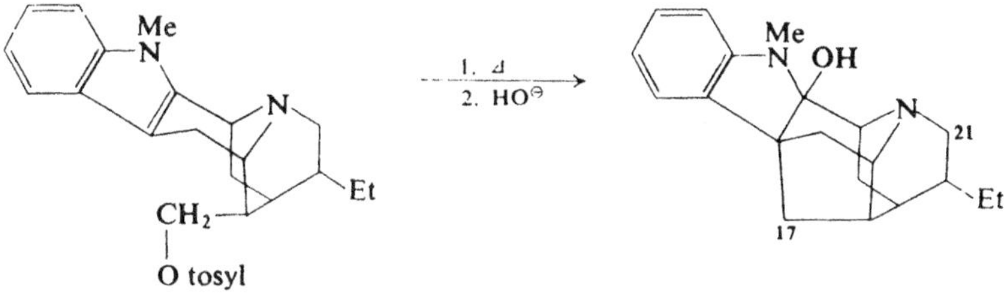

Two congeners of ajmaline merit comment. Vomilenine a potential αβ-unsaturated aldehyde was readily convertible into perakine (Michael addition of an amine to an αβ-unsaturated aldehyde) by brief reflux in acetic acid. It is well possible that perakine is an artefact.

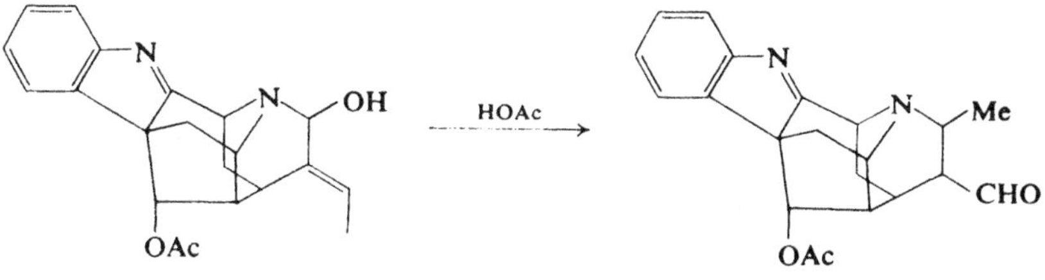

vomilenine perakine

Sarpagine. The structure of sarpagine was first put forward from a limited amount of data and biogenetic speculation. The correctness of this guess was established by its conversion by unexceptional methods into 1-methyl-10-deoxydihydrosarpagine identical in all respects with 21-deoxyajmalol-B. So far 16-epi-sarpagine has not been recognized as a natural product.

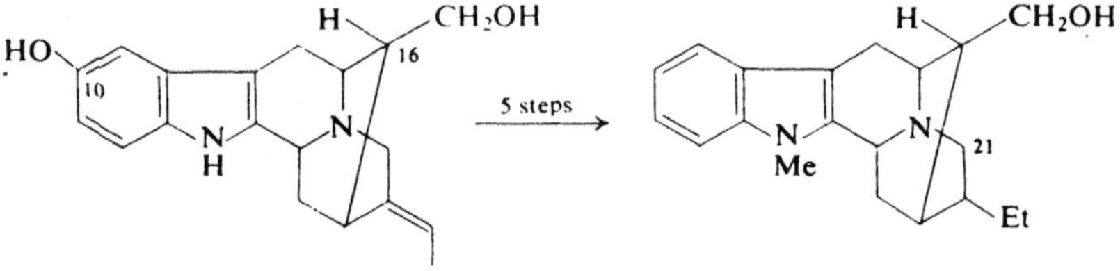

sarpagine 21-deoxyajmalol-B

Vobasine. This alkaloid with an ultraviolet spectrum indicative of a 2-acylindole moiety is one of about 16 congeners, the continued study of which may bring about a better understanding of indole alkaloid biosynthesis. They could be biosynthesized by a ring opening of their pentacyclic sarpagine-like counterparts but the reverse may well be true. To carry the last suggestion one stage further some known quaternary alkaloids might be artefacts which result from the attempted isolation of their tertiary ten membered ring equivalents, where a hydroxyl replaces the 2-keto function. In the particular case of vobasine, configuration of the system is apparently against any close approach of the basic nitrogen and the carbonyl group since the spectral shifts characteristic of

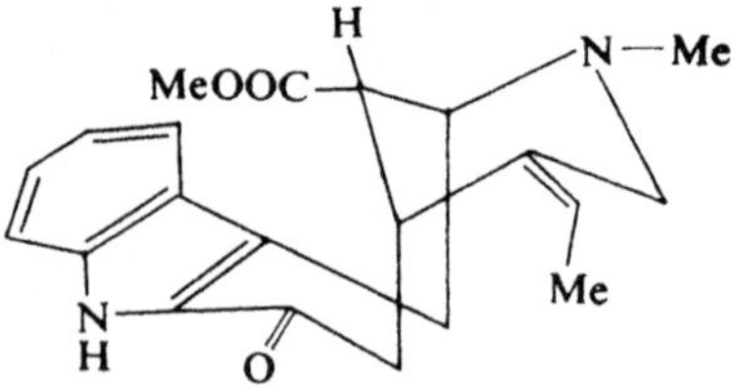

vobasine

burnamicine (p. 64) at high and low pH were not observed. Despite the lack of ring closure in polar solvents, cyclization of this type of compound has been accomplished in the following way. Perivine (*nor*vobasine) was reduced with sodium borohydride in aqueous methanol saturated with carbon dioxide. The resultant alcohol in refluxing xylene gave the cyclized product, which after epimerization (base) and lithium aluminium hydride reduction furnished 10-deoxysarpagine.

Dihydrovobasine is a very reactive compound in acidic solution and in the presence of a suitable nucleophile it functions as an alkylating agent. Thus it coupled with voacangine to furnish the bisindole alkaloid voacamine (Chart 8.4). Since this reaction is reversed in strongly acidic solution (i.e. the degradative method), partial synthesis has become the preferred route for solving the structures of the remaining members of this dimeric set. This work was made possible and backed up at all times by all the modern physical methods. Of particular interest was the way that the relative stereochemistry of the 16-methoxycarbonyl group of vobasine was elucidated. It turned out that its methyl was very powerfully shielded by the aromatic system and/or the carbonyl group, whereas in isovobasine (C-16 epimer) the signal was in the normal position.

Voachalotine. Voachalotine can be regarded as a trapped intermediate along the way of ajmaline biosynthesis. The presence of the methoxycarbonyl group confers some interesting reactions on the molecule. Attempted Oppenauer oxidation gave only the dehydroxymethyl compound (retro-aldolization) as also did pyrolysis, e.g. during vapour phase chromatography or attempted volatilization into a mass spectrometer. The hydroxymethyl because of this facile process is also the source of a spurious N-methyl in the Herzig-Meyer estimation.

The action of cyanogen bromide on the alkaloid afforded the cyclic ether whose formation is sketched in Chart 8.5. Chromic acid oxidation gave not the expected aldehyde but instead a cyclic ether. Its formation may be mechanistically related to the course

CHART 8.4

VOBASINE AND ITS CONVERSION INTO VOACAMINE

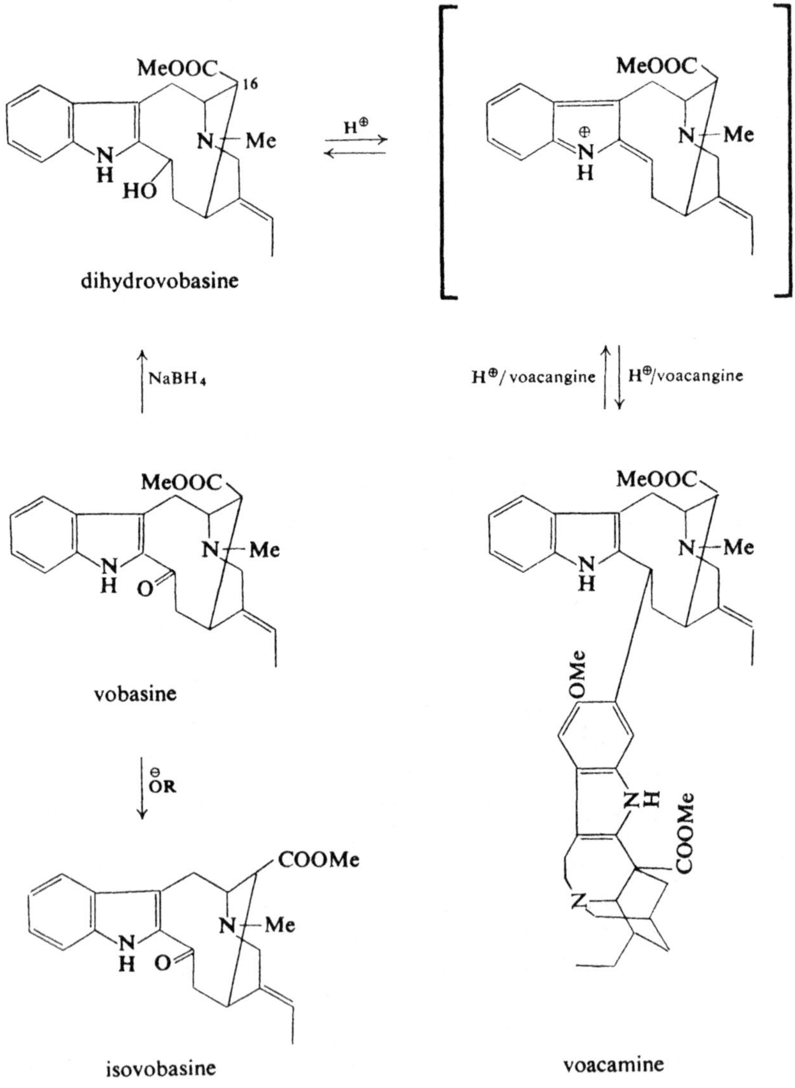

CHART 8.5

PROPERTIES OF VOACHALOTINE AND THE LEAD TETRAACETATE OXIDATION OF 21-DEOXYAJMALOL-A

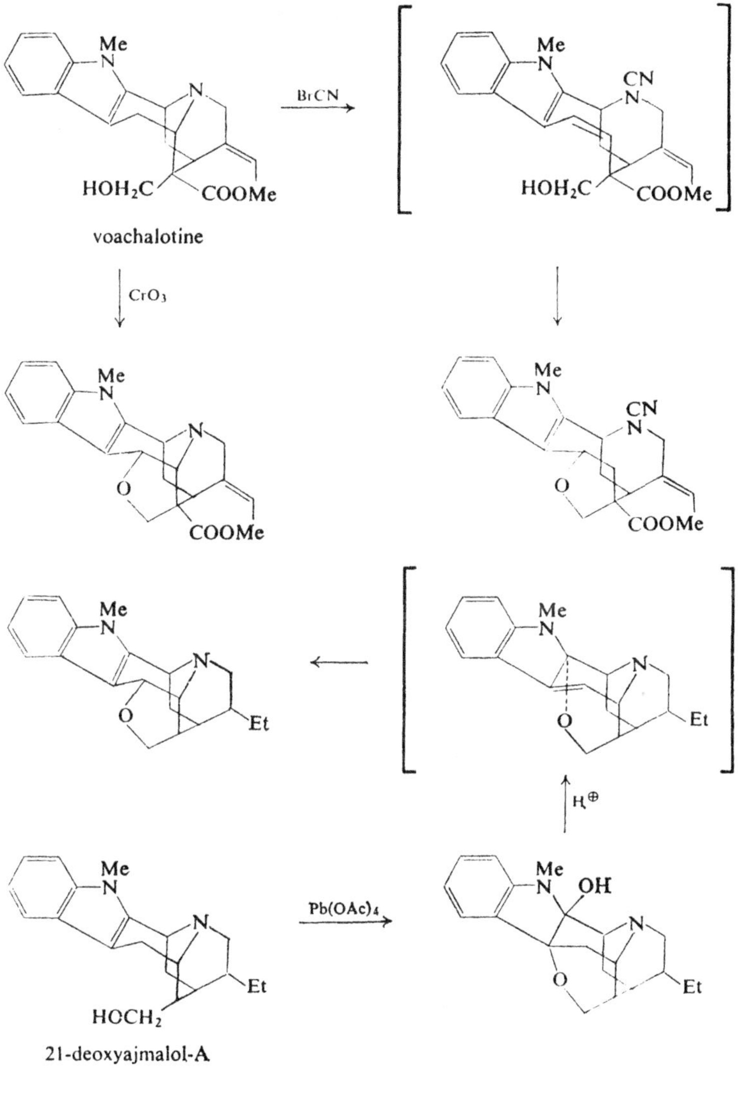

of the lead tetracetate oxidation carried out on 21-deoxyajmalol-A. The primary product, of this oxidation rearranged upon heating in dilute sulphuric acid to an analogous tetrahydrofuran. A plausible course for these transformations is given in Chart 8.5.

SUGGESTED READING

ANET, F. A. L., CHAKRAVARTI, D., ROBINSON, R. and SCHLITTLER, E., Ajmaline Part I, *J. Chem. Soc.*, **1954**, 1242.

BUECHI, G., MANNING, R. E. and MONTI, S. A., Voacamine and voacorine, *J. Amer. Chem. Soc.*, **86**, 4631 (1964).

CHAPTER IX

THE IBOGA ALKALOIDS

THE alkaloids of this group are derivatives of Type II precursor (Chart 2.1) and were first isolated from the African drug plant, *Tabernanthe iboga* hence their class name. They are substitution or simple transformation products of the parent pentacycle, ibogamine or its 16,21 *seco* derivative in which all or some of the following combinations are possible, methoxyl at C-10 or C-11, methoxycarbonyl at C-16, hydroxyl at C-19 or C-20, and a 15,20 double bond. The structures of these compounds depend on their interrelationships with and degradation of ibogaine as well as by two X-ray proofs of structure (ibogaine and cleavamine).

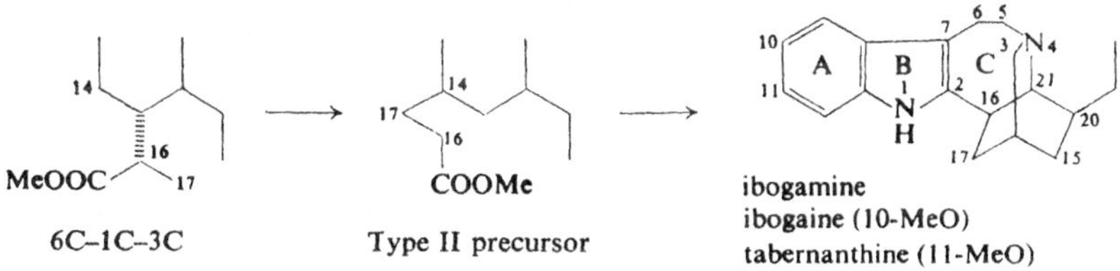

ibogamine
ibogaine (10-MeO)
tabernanthine (11-MeO)
voacangine (10-MeO, 16-COOMe)

The iboga alkaloids distinguish themselves from most of Type I compounds by the ease with which they autoöxidize to yield hydroperoxy- and hydroxyindolenines (Chart 9.1) whose further degradation products are 4-hydroxyquinolines, pseudoindoxyls and oxindoles (cf. Chapter III). Therefore, the isolation of these products from the plant cannot by itself be taken as proof of their natural occurrence. Why some of these bases are so susceptible to oxygen is not known and is certainly not so obvious as the release of bond strain which accompanies the conversion of mavacurine (a Type I alkaloid) into fluorocurine via alkaloid-Y.

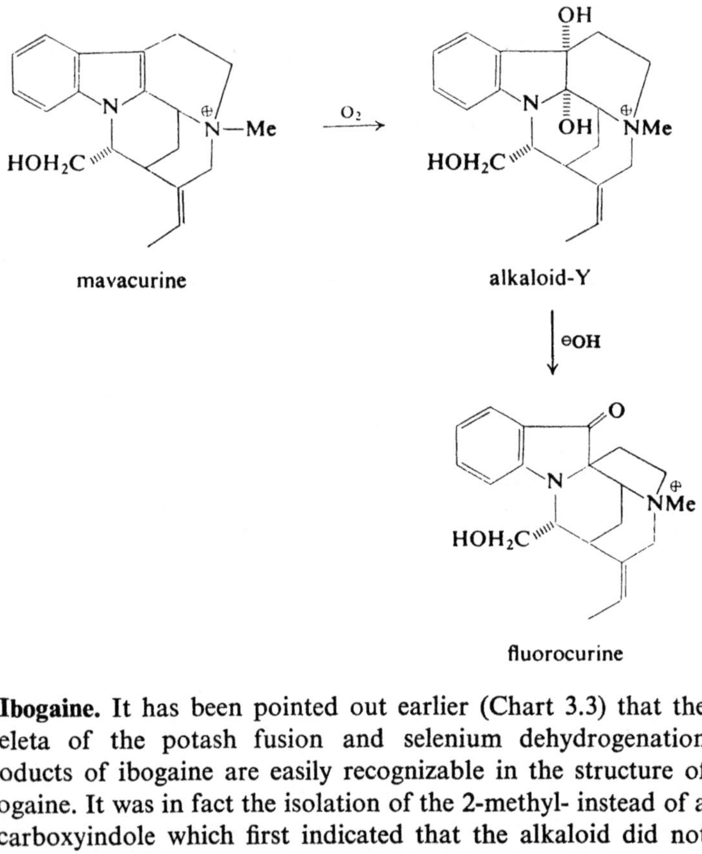

Ibogaine. It has been pointed out earlier (Chart 3.3) that the skeleta of the potash fusion and selenium dehydrogenation products of ibogaine are easily recognizable in the structure of ibogaine. It was in fact the isolation of the 2-methyl- instead of a 2-carboxyindole which first indicated that the alkaloid did not contain a tetrahydro-β-carboline moiety and therefore had to have a novel constitution. The presence of at least an unblocked tetrahydro-β-carboline system was also excluded on the grounds that no extension of conjugation was observed spectroscopically when ibogaine was treated with either *t*-butyl hypochlorite or lead tetraacetate. A five-membered ring C was also eliminated because autoöxidized ibogaine could be converted into a pseudoindoxyl (iboluteine), a 4-quinolinol (iboquine) and the case of tabernanthine (MeO at C-11) an oxindole as well, all rearrangements which involve a ring C contraction. Note that the formation of iboquine established the presence of the C-6 methylene group. Yet another property which distinguished it from many alkaloid classes was

CHART 9.1

OXIDATIVE TRANSFORMATIONS OF IBOGAINE

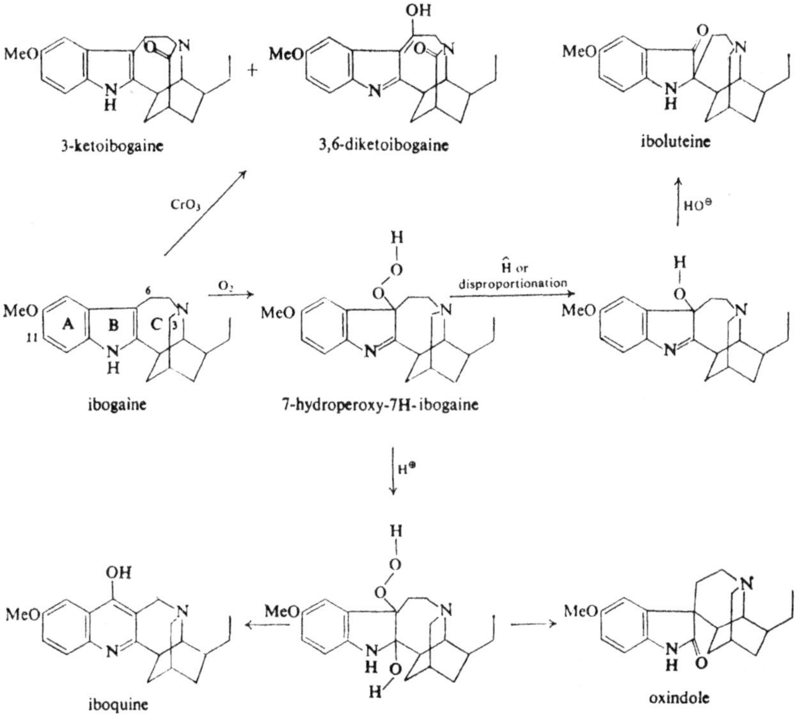

the chromic acid oxidation to a lactam (3-ketoibogaine) and to 3,6-diketoibogaine (possibly oxidative attack at C-7 followed by re-arrangement cf. 21-deoxyajmalol-A cyclic ether, Chart 8.5). All these transformations are sketched in chart 9.1.

The availability of iboluteine made possible a proof of structure

CHART 9.2

DEGRADATION OF IBOLUTEINE

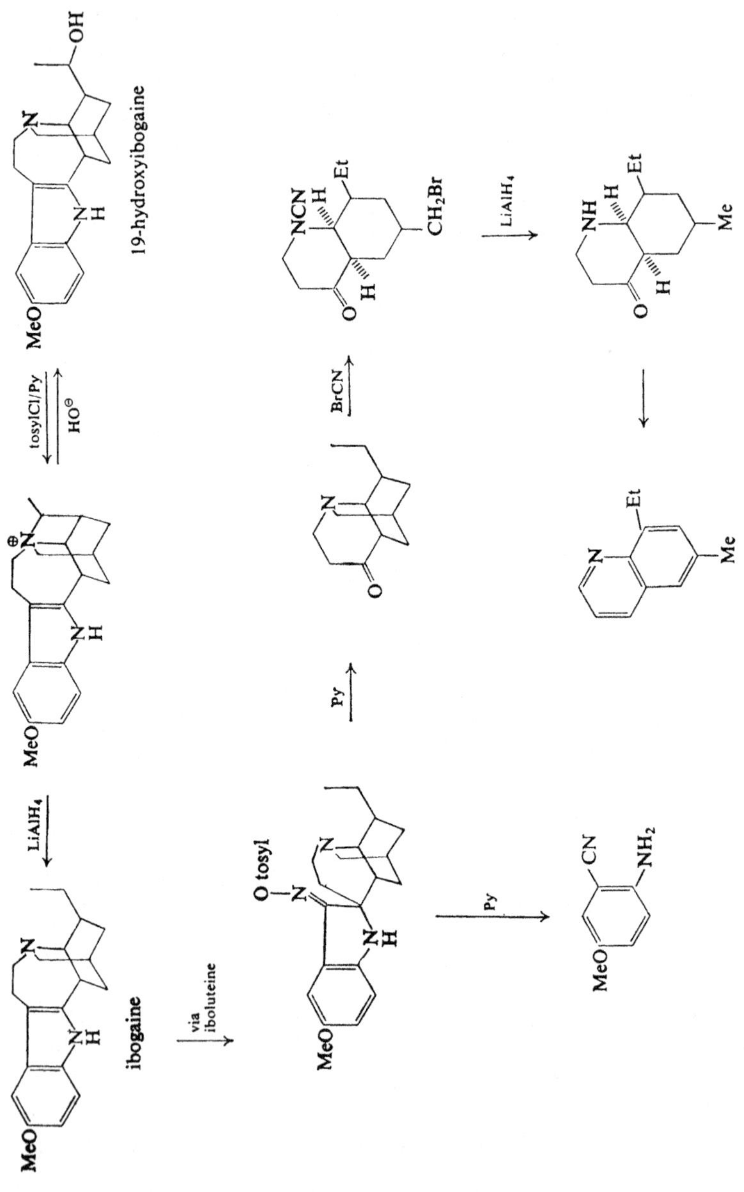

along with an unequivocal correlation with ibogamine and tabernanthine (Chart 9.2). When O-tosyliboluteine oxime was refluxed in pyridine it underwent a second order Beckmann rearrangement (cf. preparation of the Wieland-Gumlich aldehyde, Chart 6.2) to provide 4-methoxyanthranilonitrile and the ring C contracted ketone. The same ketone and the corresponding anthranilonitriles were obtained from ibogamine and tabernanthine. The amino ketone subjected to a von Braun cyanogen bromide reaction gave an N-cyano ketone convertible into 6-methyl-8-ethylquinoline. The rotatory dispersion of this ketone was compared with a number of decalones and found to resemble most closely *l*-*cis*-9-methyl-4-decalone whose absolute configuration was known.

CHART 9.3
VON BRAUN DEGRADATION OF IBOGAINE

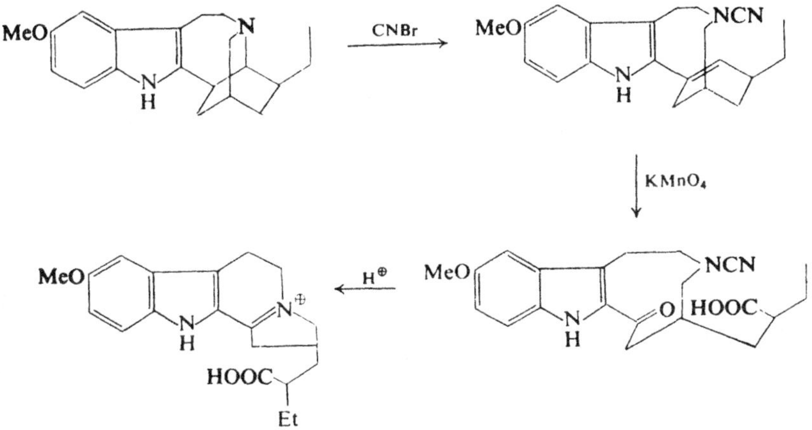

The configuration of the ethyl group as well as an independent proof of the structure came from a three dimensional X-ray analysis. The stereochemistry of the ethyl was also a consequence of the properties of 19-hydroxyibogaine (iboxygaine or kimvuline). Upon treatment with tosylchloride in pyridine a quaternary salt was formed reconverted into the alcohol again by aqueous sodium

CHART 9.4

PROPERTIES OF VOACANGINE

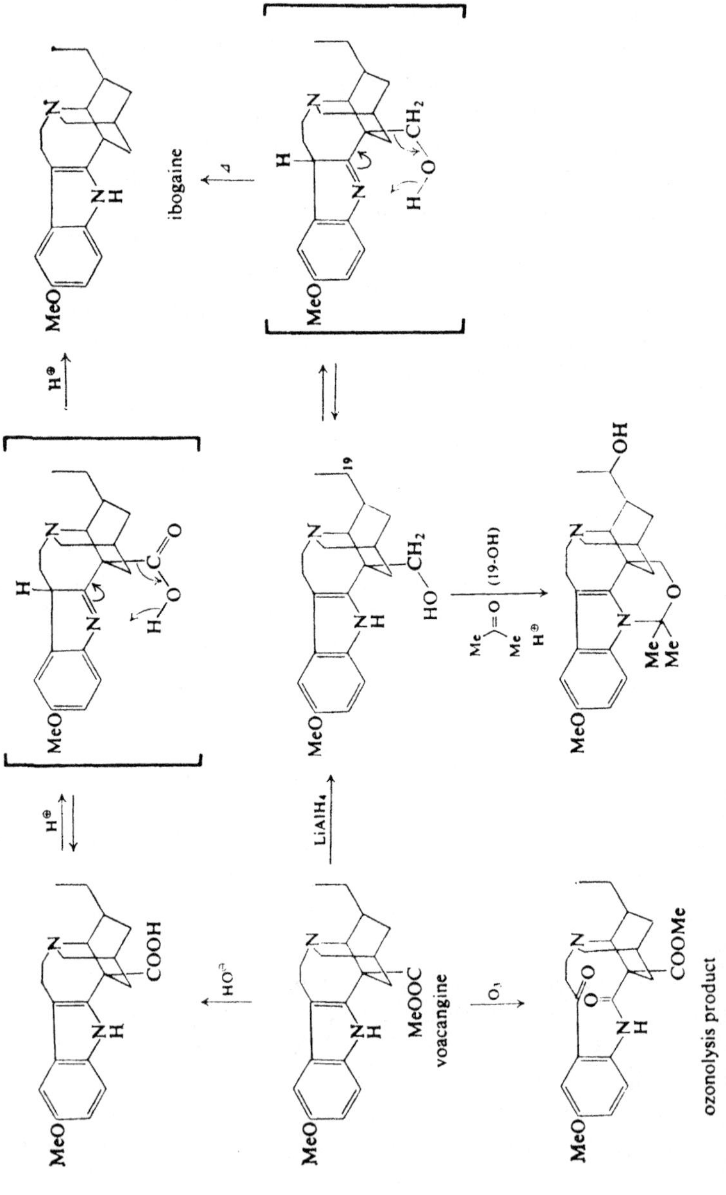

hydroxide. Reduction of this quaternary salt with lithium aluminium hydride gave ibogaine (Chart 9.2).

Another degradation of ibogaine although not carried out as far as a known compound is of interest because it led to a β-carboline (Chart 9.3).

Voacangine. The properties of voacangine are representative of the 16-methoxycarbonylibogamines. Voacangine is a weak base some 2 pK_a units less than ibogaine for no very obvious reason. Voacangic acid was readily decarboxylated thermally or by reflux in mineral acid (cf. β-ketoacids). Voacanginol (lithium aluminium hydride reduction product of voacangine) eliminated formaldehyde above its melting point to furnish ibogaine (cf. a retroaldolization). These eliminations probably proceed via the 7H-indolenine form (nitrogen isostere of a ketone) and are facilitated by the almost planar geometry of the aromatic rings and the C-16 substituent. The site of the carboxyl group was confirmed by treating 19-hydroxyvoacanginol with acetone containing hydrogen chloride to afford the acetonyl derivative. Ozonolysis of voacangine gave a complex acylamino acetophenone which did not cyclize to the corresponding 4-hydroxyquinolinol with base probably because of the unfavourable configuration of the ten-membered ring (cf. vobasine). It is for this reason that an acylaminoacetophenone intermediate was not invoked in the conversion of 7-hydroperoxy-7H-ibogaine into iboquine (Chart 9.1).

Catharanthine and Cleavamine. The occurrence of catharanthine is presently confined to *Vinca rosea* and an understanding of its chemistry was of great value in developing the structure of the antileukemic drug, vincaleukoblastine. The presence of the 15,20 double bond in catharanthine allowed a facile palladium catalysed dehydrogenation to take place at a low temperature (~ 160°). This reaction may have proceeded via the intermediacy of a retro Diels–Alder product before yielding 3-ethylpyridine. When the double bond was removed a higher temperature was necessary (~250°) and 3-methyl-5-ethylpyridine was the isolated base.

CHART 9.5

PROPERTIES OF CATHARANTHINE

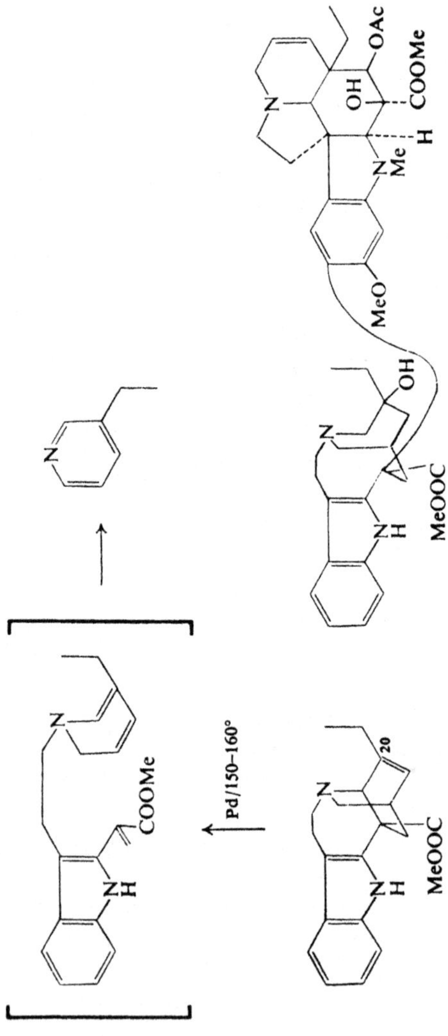

catharanthine

vincaleukoblastine

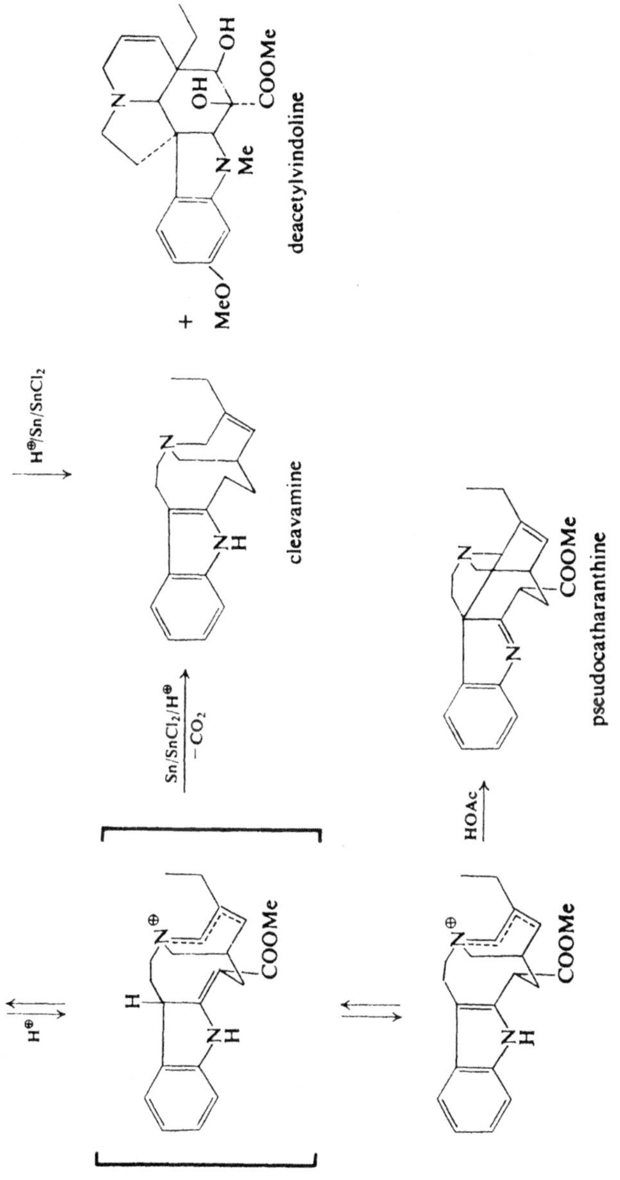
H⊕
Sn/SnCl2/H⊕
−CO2
HOAc
H⊕/Sn/SnCl2
COOMe
cleavamine
pseudocatharanthine
deacetylvindoline
MeO
OH
Me
+

Both catharanthine and vincaleukoblastine in concentrated hydrochloric acid plus a reducing agent (stannous chloride) gave cleavamine (when first isolated the structure of the "cleaved"

CHART 9.6

EPIMERIZATION OF IBOGAMINE AND THE OXIDATION OF 16-METHOXYCARBONYLDIHYDROCLEAVAMINE

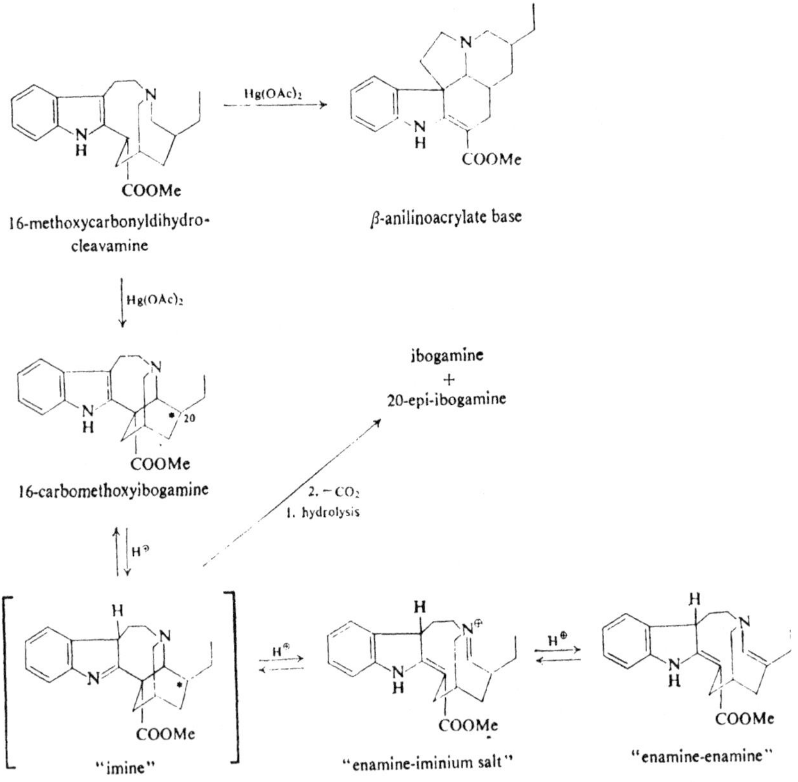

product was not known hence (?) its name). The structure and absolute stereochemistry of cleavamine was established independently of chemical considerations by the X-ray method and is in complete agreement with the rest of the iboga alkaloid work.

These reactions mediated by strong acids are a special case of the enamine chemistry of indoles outlined in Chart 3.5 and previously considered in a different structural environment for demethoxycarbonylakuammicine (Chart 6.7). In catharanthine the 15,20 double bond and the 16-methoxycarbonyl group help to stabilize the tetracyclic iminium intermediate. The products obtained depend upon the rate of decarboxylation of the pentacyclic compound as compared with reduction of the tetracycle. Even in the absence of the olefinic bond the existence of the 16,21 *seco* iminium salt must be postulated since prolonged reflux in concentrated hydrochloric acid of either 16-methoxycarbonylibogamine or its 20-epimer produced a mixture of ibogamine and 20-*epi*-ibogamine (Chart 9.6). Under gentler but non-aqueous conditions, catharanthine upon prolonged reflux in acetic acid gives rise to yet another theoretically predictable product, pseudocatharanthine (Chart 9.5).

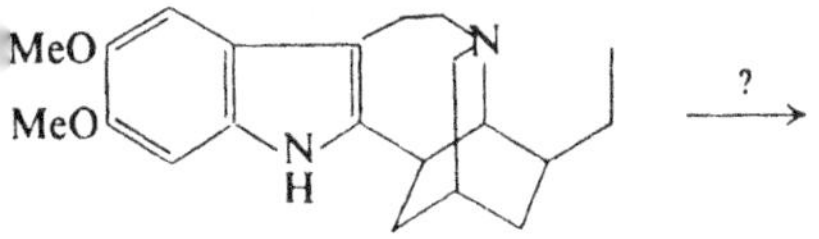

ibogaline

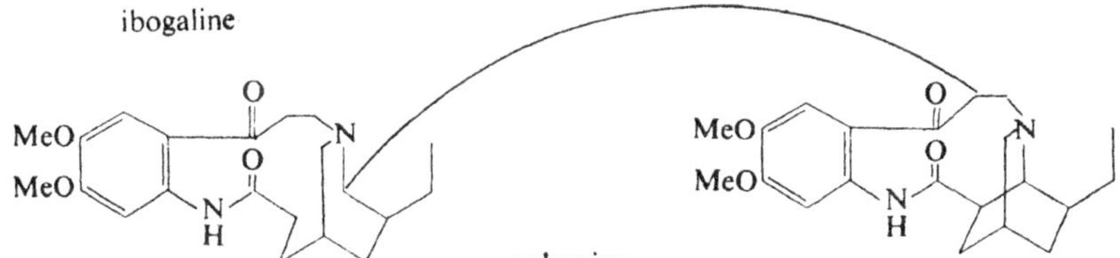

gabonine

The formation of a 16,21 *seco* compound is also involved in the genesis of gabonine a minor dimeric base of *Tabernanthe iboga* which no longer contains indole nuclei but rather acylamino-acetophenone chromophoric moieties and could well be an artefact derived from the autoöxidation and rearrangement of ibogaline (10,11-dimethoxyibogamine).

From the discussion above it becomes easier to understand the problems involved in attempting a synthesis (cf. voacamine,

Chart 8.4) of vincaleukoblastine from vindoline and the iboga monomeric units presently available. The problem is important because of the low yield (artefact?) of vincaleukoblastine from *Vinca rosea* and the possibility of preparing a better drug synthetically. Finally it might even be asked if the isolated iboga alkaloids are natural products since they could all have been formed by ring closure of labile compounds equivalent to the 16,21 *seco* iminium salts? It has in fact been shown that mercuric acetate oxidation of 16-methoxycarbonyldihydrocleavamine gave not the ring opened products but instead the corresponding pentacycles, a β-anilinoacrylate base and both the 20-epimeric-16-carbomethoxy ibogamines (Chart 9.6).

CHAPTER X

THE EBURNAMINE ALKALOIDS

IN THIS and the next chapter the chemistry of Type III precursor alkaloids will be discussed. They form two distinct groups, the indole (eburnamine) and indolines (aspidospermine), (Chart 10.1) which have their equivalents in the Type I groups as well as new variants illustrated by pleiocarpine and schizozygine. The facile ring opening and closing of the aspidospermine-quebrachamine pair has an exact parallel in the chemistry of demethoxycarbonylakuammicine (Chart 6.7). Unlike Type I alkaloids these bases have no stereochemical invariant centres and *d*-, *l*- and *rac*- forms are known. Although the simpler members have been synthesized the hexa- and heptacyclic congeners are still unsolved challenges. No interconversion between the eburnamine and the aspidospermine compounds has yet been accomplished.

The development of the chemistry of the eburnamine alkaloids all from apocynaceous plants was greatly facilitated by the knowledge which had been accumulated on the canthin-6-ones, bases originally isolated from *Pentaceras australis* (*Rutaceae*) and more recently from simarubaceous plants. For this reason a short discussion of their chemistry is in order.

Canthin-6-ones. From *Pentaceras australis*, three optically inactive coloured alkaloids were isolated. The structure of canthin-6-one followed from its analytical data, physical properties and potassium permanganate oxidation to β-carboline-1-carboxylic acid and synthesis from 5-methoxycanthin-6-one (*vide infra*). (Chart 10.2).

CHART 10.1

BIOGENETIC RELATIONSHIPS BETWEEN REPRESENTATIVE TYPE III INDOLE ALKALOIDS

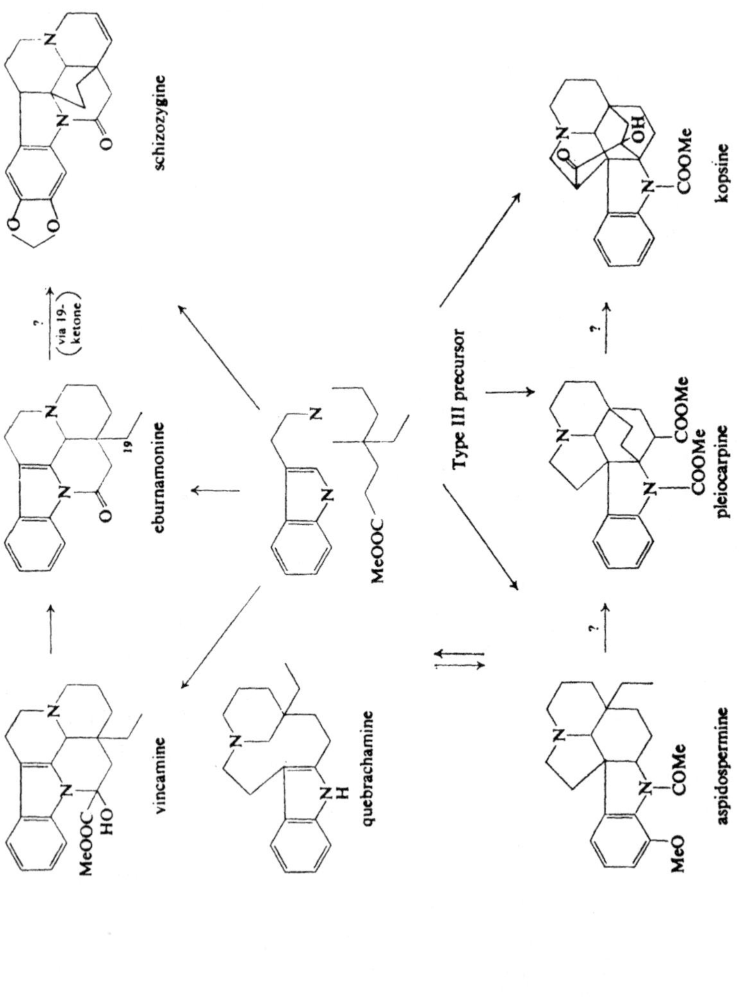

CHART 10.2
PROPERTIES OF THE CANTHIN-6-ONES

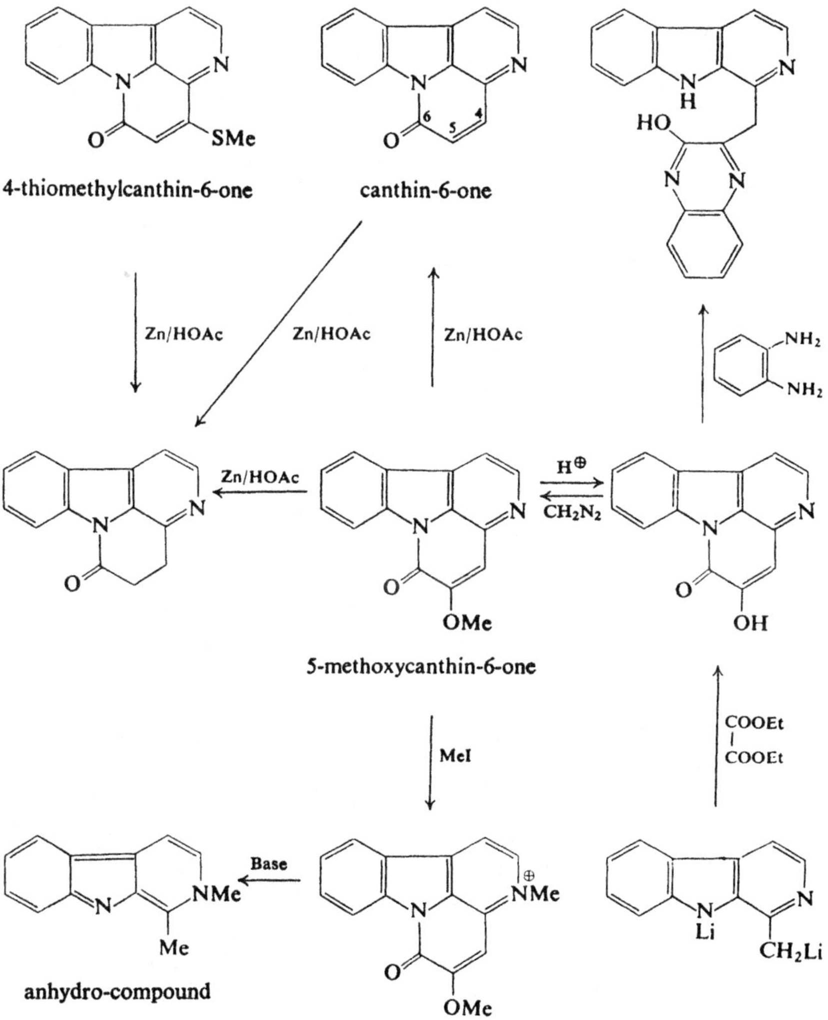

A second alkaloid 5-methoxycanthin-6-one was demethylated with hydrobromic acid in acetic acid to a phenol which formed water insoluble alkali salts. Upon treatment of the phenol with diazomethane in methylene chloride or its O-acetate in moist ether the alkaloid was regenerated. Like canthin-6-one, potassium permanganate oxidation gave β-carboline-1-carboxylic acid and the phenol with *o*-phenylenediamine furnished a hydroxyquinoxaline. These facts established its structure. Reduction of the alkaloid with zinc in acetic acid resulted in a mixture of canthin-6-one and its 4,5-dihydro-derivative. The latter derivative could be easily dehydrogenated to the former with sulphur or selenium.

A synthesis of 5-hydroxycanthin-6-one was accomplished by condensing the dilithium derivative of 1-methyl-β-carboline with diethyloxalate. A formal reversal of this synthesis was brought about by heating 5-methoxy-canthin-6-one methiodide in base whereupon anhydro-1,2-dimethyl-β-carboline crystallized out.

The third alkaloid from the Australian plant turned out to be 4-methylthiocanthin-6-one. Its chemistry paralleled that of the above bases. Oxidation yielded β-carboline-1-carboxylic acid, reduction gave canthin-6-one. It was synthesized as follows: β-carboline-1-carboxylic acid chloride was condensed with the magnesium ethoxy derivative of malonic ester to yield after hydrolysis 4-hydroxycanthin-6-one. This compound first treated with phosphorous oxychloride then reacted with potassium methyl mercaptide gave the alkaloid.

Eburnamonine. Among the alkaloids of *Hunteria eburnea* were four interrelated bases. Two of them eburnamine and isoeburnamine were diastereoisomeric pentacyclic indoles convertible by acids into eburnamenine, an N-vinylindole, on the one hand, and by chromic acid into eburnamonine, an N-acylindole, on the other. Reduction of eburnamonine with lithium aluminium hydride regenerated the alcohols. When eburnamonine was heated with selenium for a few minutes it gave 4-ethyl-4-propyl-4,5-dihydrocanthin-6-one. Prolonged heating eventually gave 4-ethyl- and 4-propylcanthin-6-ones, which were synthesized by condensa-

CHART 10.3

SOME PROPERTIES OF EBURNAMONINE AND EBURNAMINE

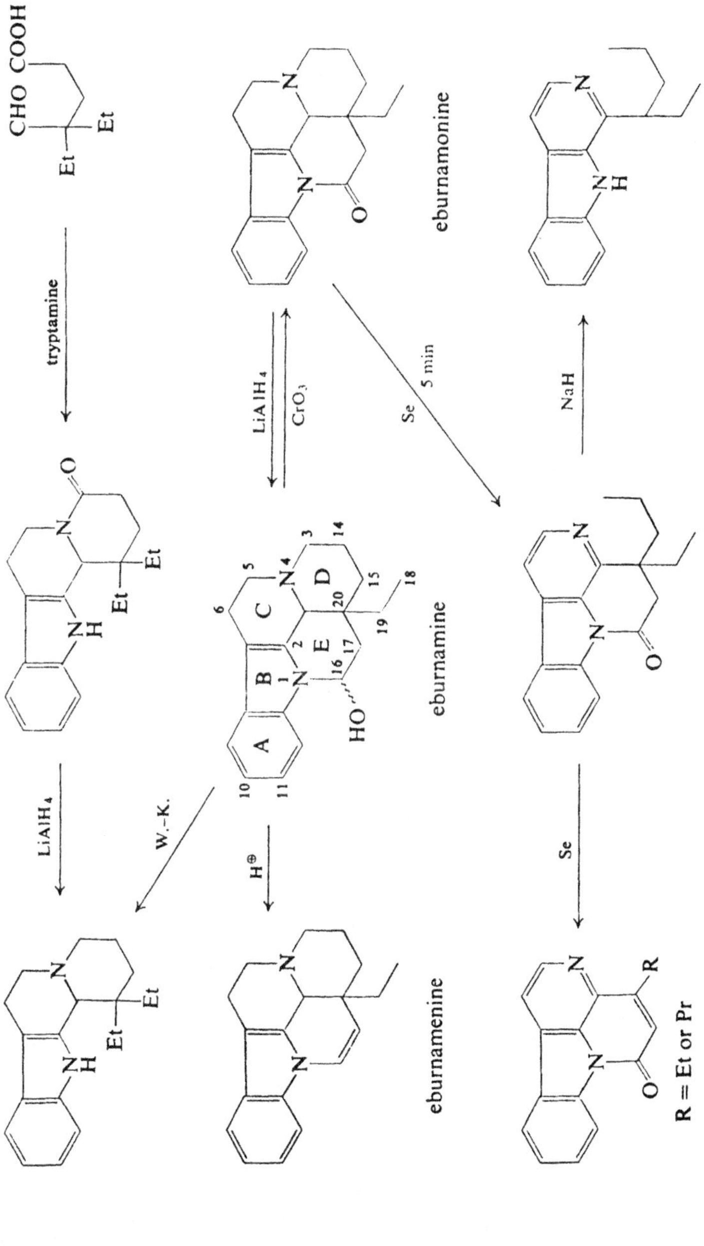

CHART 10.4

SYNTHESIS OF EBURNAMONINE

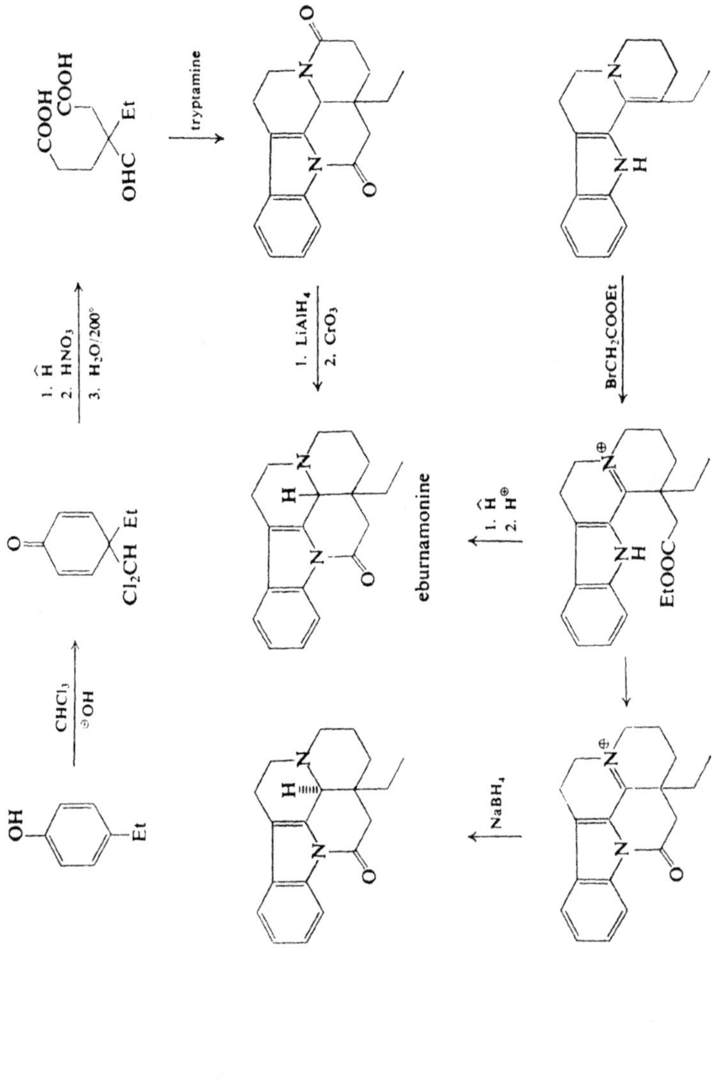

tion of diethyl oxalate with the dilithium derivative of the appropriate 1-alkyl-β-carboline followed by reductive removal of the 5-hydroxy group of the product. When 4-ethyl-4-propyl-4, 5-dihydrocanthin-6-one was subjected to prolonged reflux with sodium hydride in toluene, it eliminated the acetyl residue to form *l*-1-(1-ethylbutyl)-β-carboline (Chart 10.3).

These results did not unequivocally establish the structure of eburnamonine which was secured not only by a total synthesis but also by a degradation of eburnamine (*vide infra*) which settled the uncertain features of ring D and the ethyl group. Condensation of β-ethyl-β-formyl-adipic acid with tryptamine gave *dl*-eburnamonine lactam which upon reduction with lithium aluminium hydride followed by oxidation of the resultant *dl*-eburnamines, afforded *dl*-eburnamonine. This synthesis gave no information concerning the stereochemistry of the CDE ring junctions but this was forthcoming from a second synthetic approach. An enamine condensation of ethyl bromoacetate with the appropriate indoloquinolizidine gave the Δ^3-compounds whose properties towards reduction indicated that in eburnamonine the DE rings are *cis* fused (Chart 10.4).

Eburnamine. Eburnamine and isoeburnamine are diastereoisomeric alcohols whose structures followed from their oxidation to eburnamonine and Wolff–Kishner reduction to *d*-1,1-diethyloctahydroindolo[2,3-a]quinolizine. The last compound was dehydrogenated (palladium catalysed hydrogen transfer to maleic acid) and reduced (sodium borohydride) to the racemic base identical with the synthetic sample (Chart 10.3).

Neither eburnamine nor isoeburnamine showed any properties, with the exception of the Wolff–Kishner reaction, which would indicate their being in equilibrium with the theoretically tautomeric aldehyde however other reactions imply the intermediacy of another tautomer, the iminium ion and the evidence for this is sketched in Chart 10.5. It is clear that because of the use of acid in the isolation of these alkaloids it is not possible to be sure if eburnamine or eburnamenine are really natural products.

CHART 10.5
INTERRELATIONSHIPS BETWEEN THE EBURNAMINES AND EBURNAMENINE

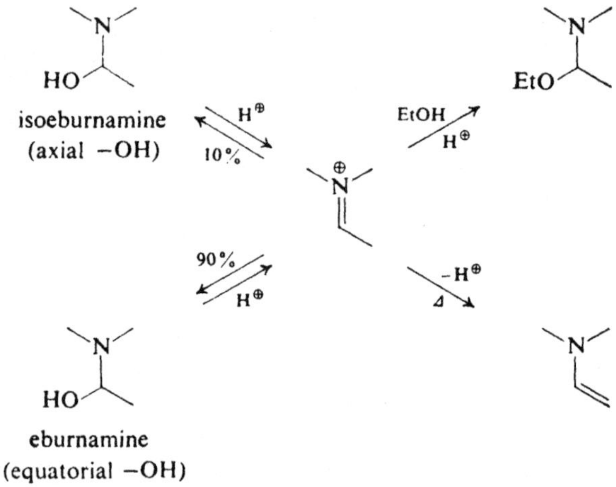

Vincamine. The principal alkaloid of this group is vincamine and so far it has only been found in the vinca species. The structure of the alkaloid became clear when it was discovered that acid treatment furnished the optical antipode of eburnamonine from *Hunteria eburnea.* Therefore as might have been expected dehydrogenation gave *l*-4-ethyl-4-propyl-4,5-dihydrocanthin-6-one. The behaviour of the hydroxyl group resembled that of the eburnamines and in an interesting experiment the iminium salt was generated by dissolving vincamine in 11*N* hydrochloric acid and detected by its characteristic ultraviolet absorption spectrum, dilution with water furnished the isolable product, apovincamine (Chart 10.6).

Other alkaloids of this group have a methoxyl group at C-11 and/or a carbonyl group at C-19. The latter type can probably be used to settle not only the entire relative stereochemistry of the molecule if a bond can be established between the ketone and the C-16 hydroxyl or methoxycarbonyl groups but also to synthesize the novel hexacyclic system of schizozygine.

CHART 10.6
PROPERTIES OF VINCAMINE

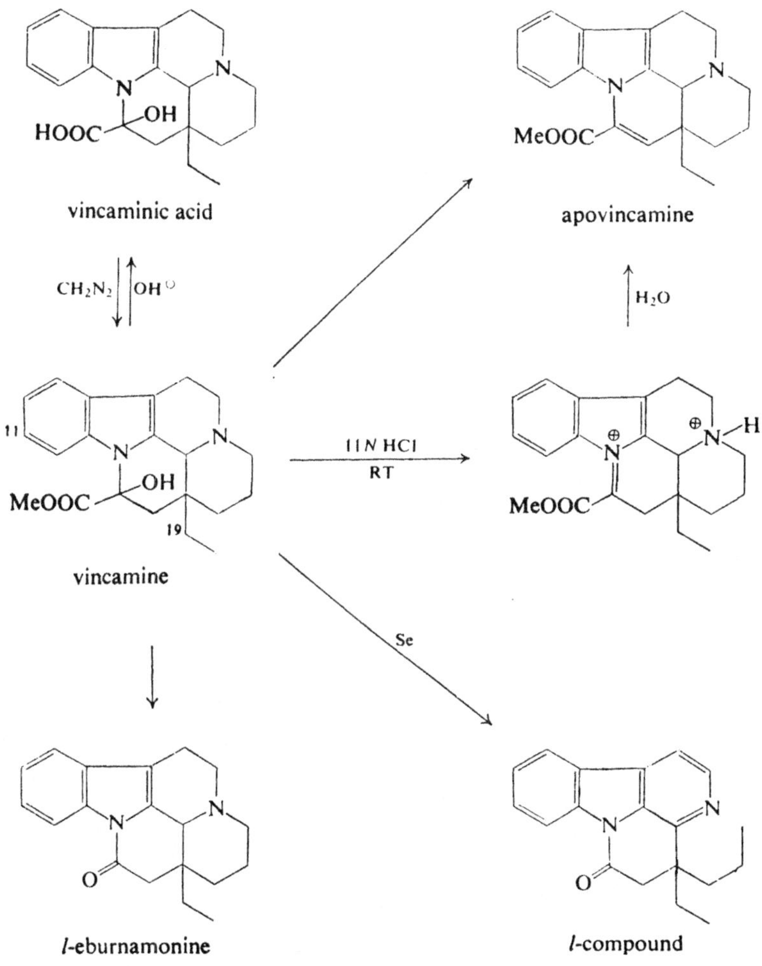

Schizozygine. This is a representative of a class of alkaloids currently found in *Schizozygia caffaeoides* whose structure rests for the present on an analysis of the physical properties of its

CHART 10.7

PROPERTIES OF SCHIZOZYGINE

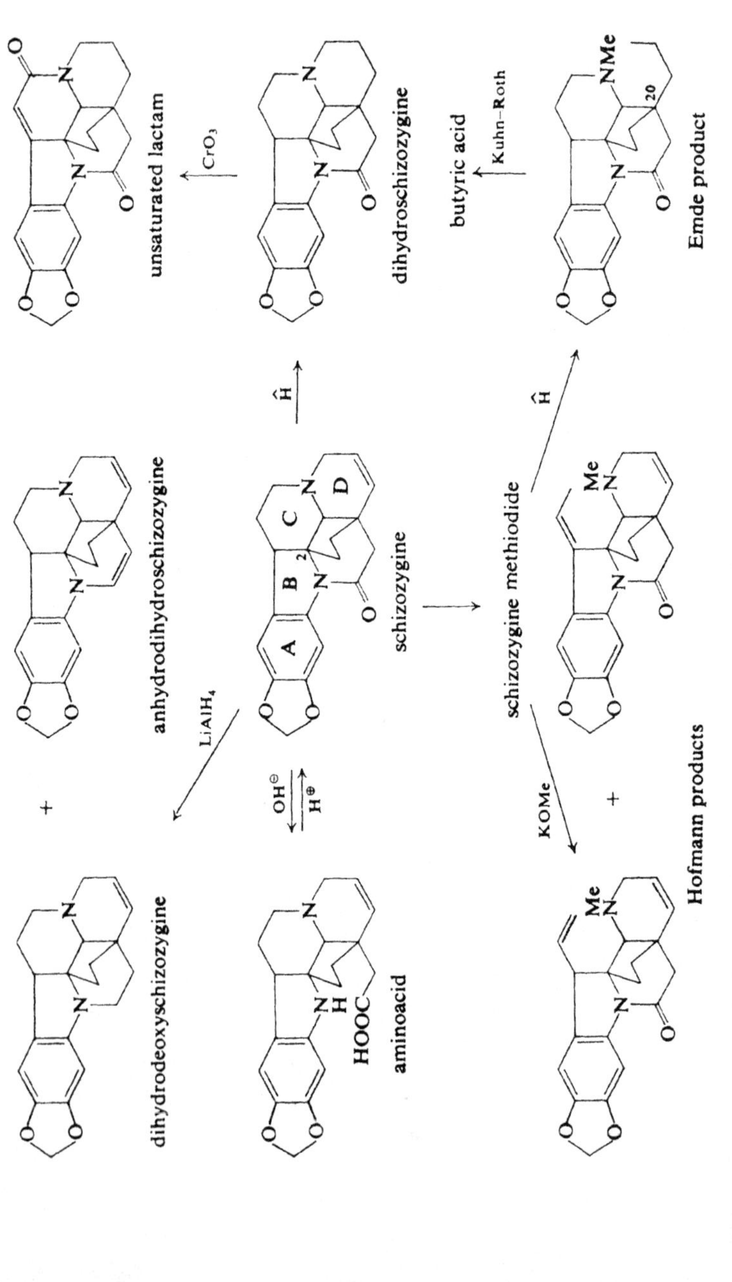

transformation products. The current formula also makes good biogenetic sense in that it is related to eburnamonine in the same way as kopsine is to aspidospermine.

Schizozygine has an ultraviolet absorption spectrum similar to brucine (dimethoxystrychnine) and like it upon lithium aluminium hydride reduction yielded an indoline and an N-vinylindoline. Hydrolysis of schizozygine gave an amino acid (indoline ultraviolet absorption spectrum) which ring closed under very mild conditions even upon treatment with diazomethane and in this property resembled eburnamoninic acid. A relationship between the basic nitrogen (incidentally a very weak base, pK_a 4·29) and the aromatic ring A was established by the obtention of an unsaturated lactam with the appropriate ultra-violet and infrared absorption properties. This conclusion was further reinforced by an examination of the Hofmann degradation products of schizozygine methiodide. In both cases it is clear that the 2 position cannot carry a proton since otherwise an indole would have resulted. The nature of ring D was illuminated by the Emde product which after hydrogenation and a Kuhn–Roth chromic acid oxidation gave butyric acid which supported the idea that C-20 was quaternary.

The other alkaloids of this group have properties which are consistent with the interpretation of the chemistry of schizozygine given in Chart 10.7, but unequivocal proofs are yet to be obtained.

SUGGESTED READING

Kuehne, M. E., The total synthesis of vincamine, *J. Amer. Chem. Soc.*, **86**, 2946 (1964).

CHAPTER XI

ASPIDOSPERMA BASES

FOR many years aspidospermine was the only member of this group. Upon drastic degradation it did not, like yohimbine, yield fragments which enabled its skeleton to be deduced and they were, in fact, as also for quebrachamine, quite misleading. Aspidospermine was never studied in the same way as strychnine probably because it was not as cheap and as available. In any event its study was taken up again in the middle fifties in a pioneering application of nuclear magnetic resonance spectra to complex natural products. Although this combination of degradative and physical work illuminated the area about the basic nitrogen the definitive structure was established by a three dimensional X-ray analysis of aspidospermine methiodide. Aspidospermine became then the reference point for other members of this group and in another pioneering investigation it became clear that its fragmentation pattern in the mass spectrum was of extraordinary usefulness for that particular pentacyclic system and this approach has been capitalized on very fruitfully in recent years. It might be as well to point out here that the mass spectral method is sensitive to stereochemistry in only special situations and by itself cannot distinguish between indole alkaloids *per se* and their analogues with inverted AB rings. This is where optical rotatory dispersion, nuclear magnetic resonance spectra and well chosen interconversions complete the story. Further examples of the wide variety of Type III precursor alkaloids isolated from aspidosperma, vinca, and pleiocarpa species, in addition to those in Chart 10.1 are given in Chart 11.1. They exist in various degrees of hydroxylation of ring A and different stereochemistries including racemates.

CHART 11.1
SOME ASPIDOSPERMA BASES

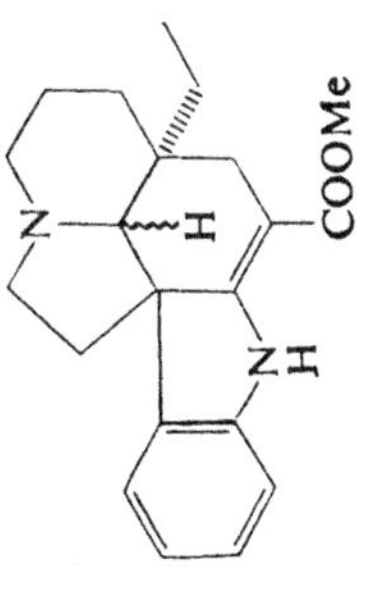

l-vincadifformine

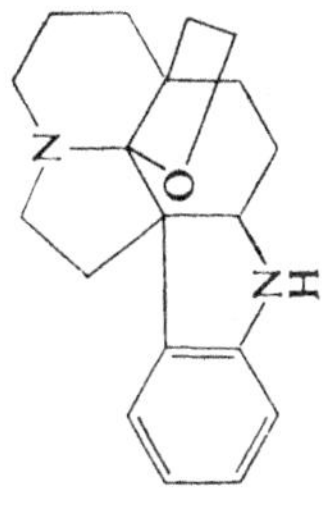

fendleridine

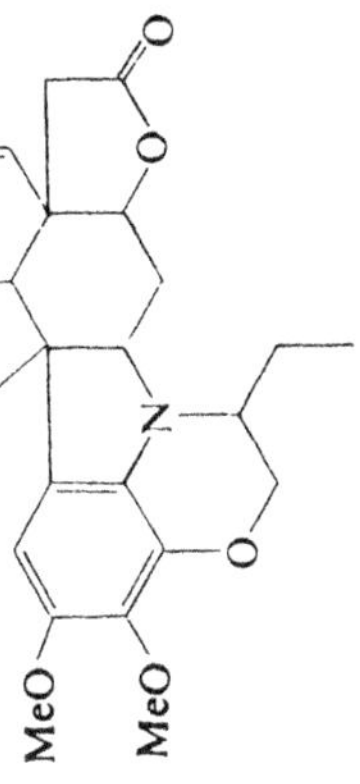

obscurinervine

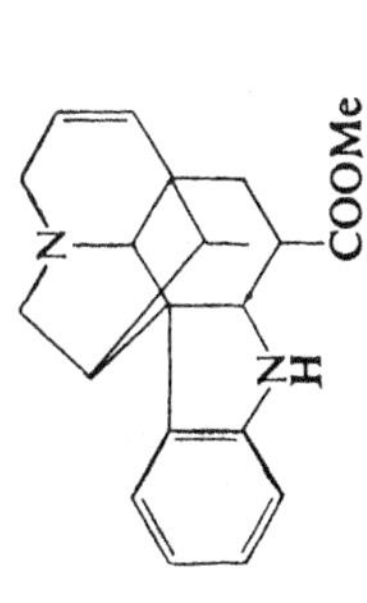

vindolinine

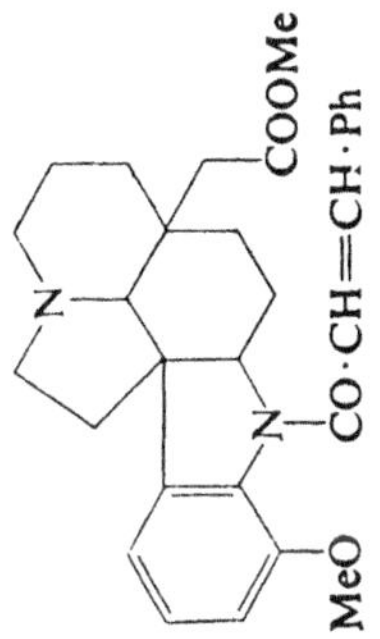

cylindrocarpine

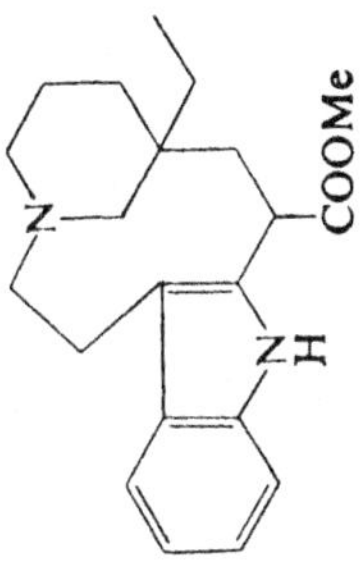

vincadine

Aspidospermine. The first inkling that the structure of aspidospermine was different from the yohimbine or strychnine types was the formation on zinc dust distillation of 3,5-diethylpyridine and since this carbon nitrogen pattern turned out not to be present in the original alkaloid it could only lead to erroneous formulae, which it did. The nature of rings A and B became clear

CHART 11.2
SOME TRANSFORMATIONS OF ASPIDOSPERMINE

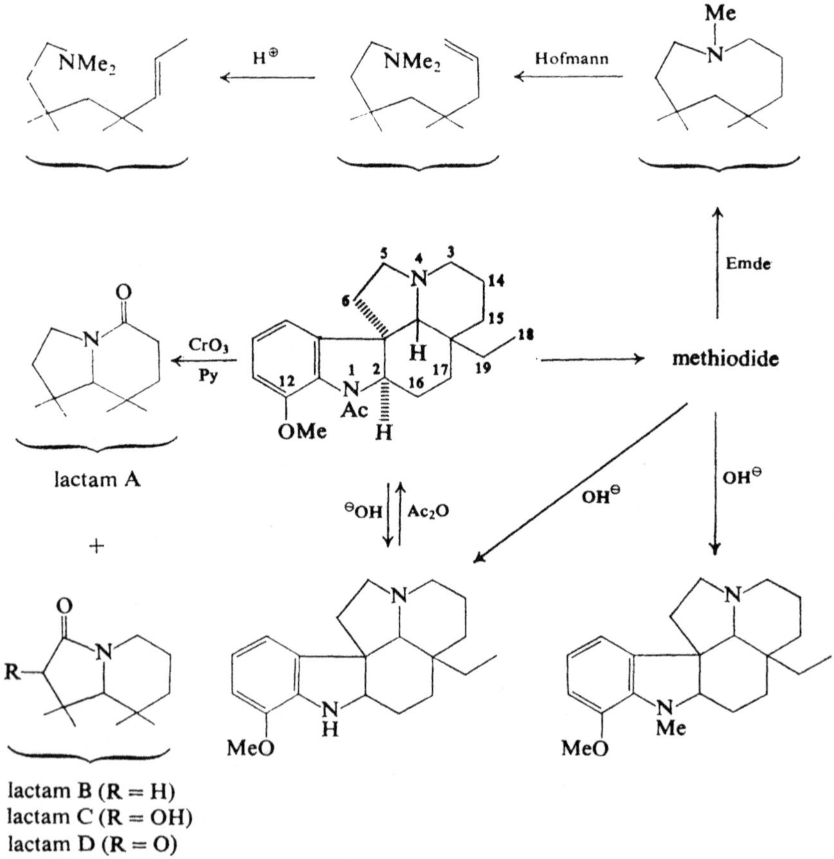

as soon as a sufficient number of model compounds became available for a comparison of ultraviolet and infrared spectra with aspidospermine and its simple transformation products. As for so many complex indole alkaloids the Hofmann degradation was not a useful reaction the products being deacetylaspidospermine and 1-methyldeacetylaspidospermine (*trans* methylation). More successful was an Emde followed by a Hofmann degradation which gave a methine whose terminal double bond was isomerized after heating in mineral acid. Chromic acid oxidation of aspidospermine gave four lactams (A→D) the most interesting being lactam D because continued work on it would have led to a chemical solution of the structural problem, however, the X-ray work removed the necessity to do so.

The first mass spectrometric investigation of desacetylaspidospermine was brought about by the desire to confirm the structure guessed for quebrachamine and this was done by converting the former compound into the 12-methoxy derivative of the latter (Chart 11.3). It was argued that if the two compounds differed only in the substitution of an aromatic proton for a methoxyl they should have identical mass spectra except for those peaks containing the aromatic residues (i.e., they would differ by 30 mass units) and this was found to be true. This comparative method is known as the mass spectrometric shift technique and has since found widespread use.

It is of value to point out here that the stereochemistry of the pentacyclic system must be the thermodvnamically stable one since reduction of the indolenine form (preferred in ether) with lithium aluminium hydride regenerated deacetylaspidospermine (cf. demethoxycarbonylakuammicine, Chart 6.7).

A total synthesis of *dl*-aspidospermine and *dl*-quebrachamine took advantage of these facts. The appropriate phenylhydrazone of the tricyclic ketone (Chart 11.3) cyclized in hot acetic acid to give the indolenine⇄indole forms, reduction of which under the correct conditions gave the desired racemic penta- or tetracyclic alkaloid. The tricyclic ketone was prepared in nine steps from butyraldehyde.

CHART 11.3

INTERCONVERSION OF PENTA- AND TETRACYCLIC ASPIDOSPERMA BASES

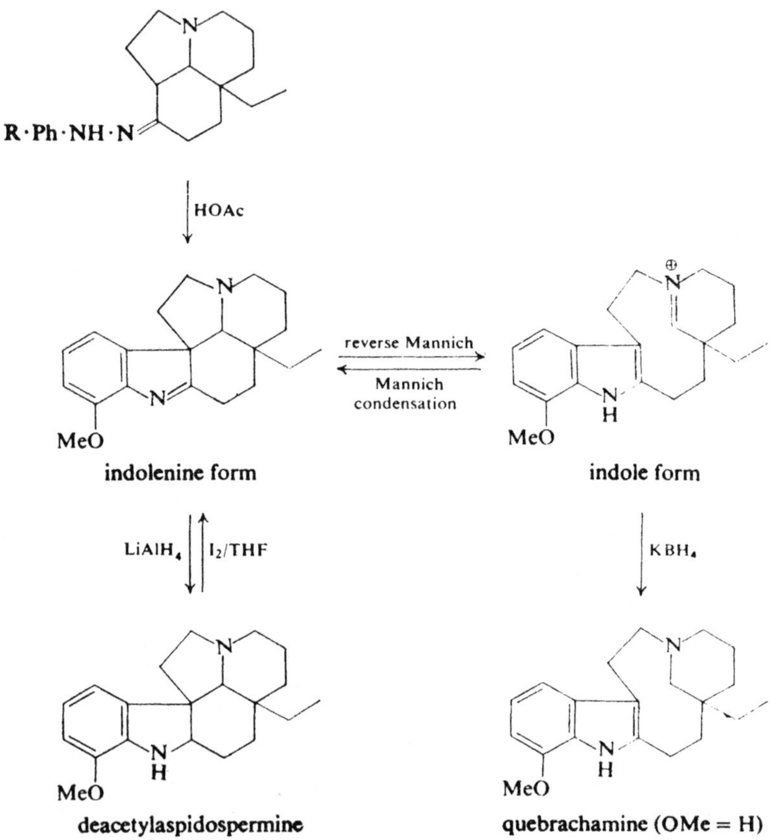

The Mass Spectrum of Aspidospermine. Upon electron impact an electron is probably kicked out of the non-bonding electrons of the alicyclic nitrogen. This ion seeks further stabilization through conversion to iminium ions either by loss of hydrogen (M-1 ion) or fission which would lead to a stabilized radical ion which in its turn may fragment. Thus aspidospermine yields species *b* (M-1),

CHART 11.4
MAJOR FRAGMENTS PRODUCED FROM ASPIDOSPERMINE UPON ELECTRON IMPACT

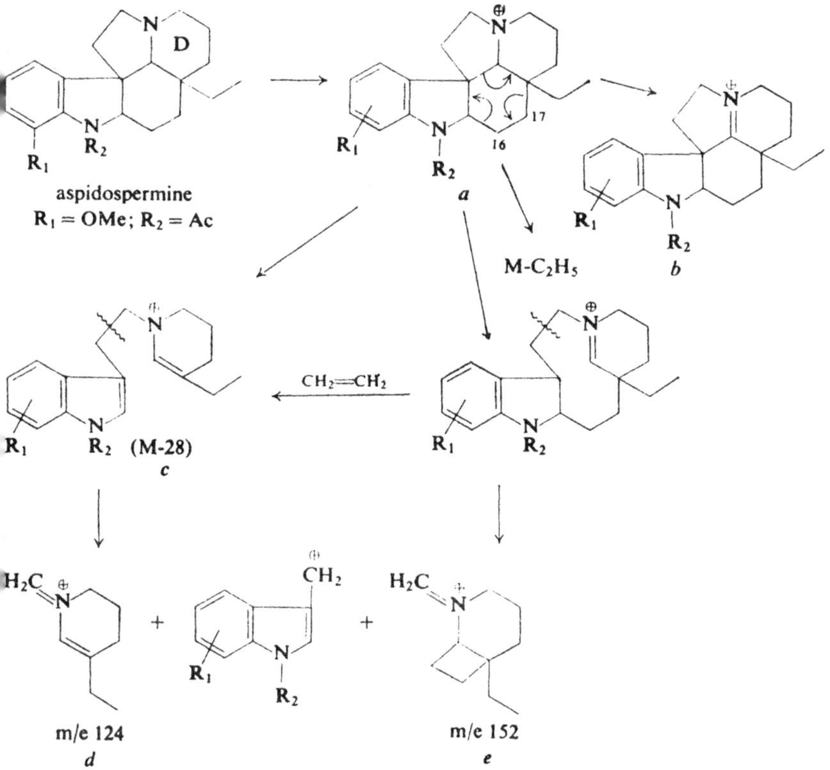

a (M)→*c* (M-28)→*d* (m/e 124) the sequence being real since the appropriate metastable peaks can be recognized. Other but less important paths involve the sequence *a*→*e* (m/e 152) and the M-C_2H_5 ion. It is easy to see that for an unknown derivative the nature of substitution if any in the aromatic system, C-16 or C-17 and ring D can be ascertained. Small peaks show up consistent for ions which have the mass of the indole fragments which must be formed by a small amount of heterocyclic cleavage. If this is coupled with a few chemical transformations as well as a sophisticated use of the mass spectrometric shift technique by the insertion of deuterium into specific locations the structure can be completely solved with the exception of the stereochemistry often with submilligram amounts of material.

Kopsine. Kopsine is a heptacyclic alkaloid which at first sight would appear to offer little hope of rapid structure elucidation, but the reader should remember how much information can be gotten out of quite simple measurements. The nature of its chromophoric moiety followed from its ultraviolet absorption spectrum; the infrared revealed its carbonyl functionalities. From the above, the elemental composition, Zeisel and Kuhn–Roth determinations and behaviour towards hydrogenation it seemed to be a heptacyclic-N-methoxycarbonylindoline containing a keto-carbonyl and a hydroxyl group.

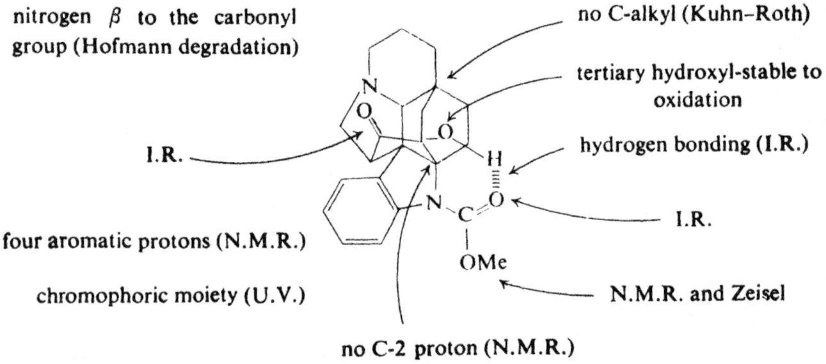

kopsine (relative stereochemistry)

The reader should also be aware that such a highly condensed rigid system lends itself (and its derivatives) to analysis by nuclear magnetic resonance spectroscopy. In this way much of the proton topography becomes known and the problem is to fit it onto a unique carbon-nitrogen matrix. One of the really important observations for kopsine was of a negative type, i.e. that there was no proton attached to C-2, in other words, C-2 must be a quaternary carbon. This fact plus the lack of a C-alkyl and the isolation of 3,5-diethylpyridine (reminiscent of aspidospermine) after zinc dust distillation led to the idea that kopsine was pleiocarpine (Chart 10.1) plus an additional ring.

The nature of this ring and its substituents and their relationships to the two nitrogens was established in part by the following experiments. The hydroxyl group was tertiary (stable to oxidation) and it was hydrogen bonded to the urethane carbonyl group. The hydroxy group was part of an α-ketal system since kopsine under a variety of conditions such as heating in an inert solvent, treatment with strong acid or base (with concomitant loss of the >N—COOMe) became equilibrated with isokopsine by an acyloin rearrangement.

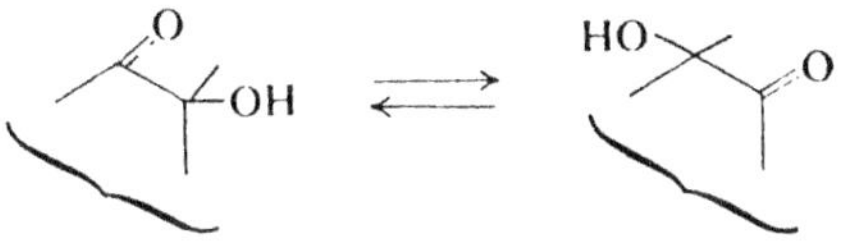

Kopsine methiodide upon treatment with sodium carbonate at room temperature gave a methine base (reverse of a Michael addition) and the methylene protons of the resulting vinylketone were not split by any other adjacent protons.

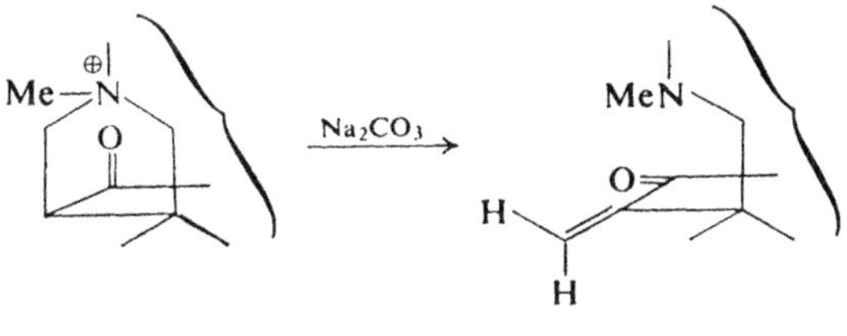

Kopsinyl iodide derived from pleiocarpine upon pyrolysis gave besides kopsinylene and other products, kopsan in 6% yield. This reaction is thought to proceed via radical intermediates. Kopsan was identical in all respects with the product obtained in very small yield by heating kopsine with hydriodic acid–red phosphorous for several days at 150°.

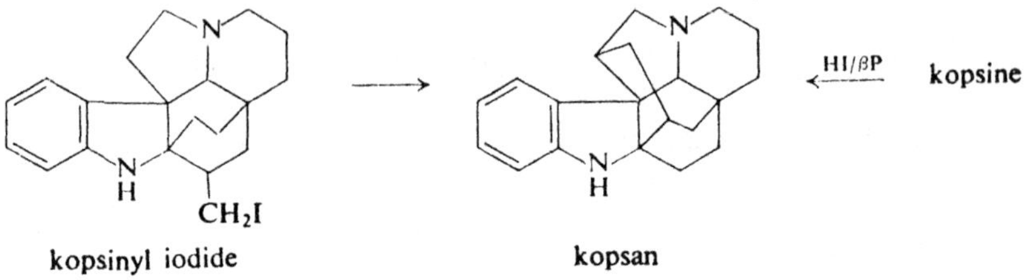

SUGGESTED READING

Biemann, K., Friedmann-Spiteller, M. and Spiteller, G., An investigation by mass spectrometry of the alkaloids of *Aspidosperma quebracho-blanco*, *Tetrahedron Letters*, **1961**, 485.

Govindachari, T. R., Pai, B. R., Rajappa, S., Viswanathan, N., Kump, W. G., Nagarajan, K. and Schmid, H., Ueber die Struktur des Kopsins, *Helv. chim. Acta*, **46**, 572 (1963).

Mills, J. F. D. and Nyburg, S. C., The molecular structure of aspidospermine, *Tetrahedron Letters*, **11**, 1 (1959).

Witkop, B. and Patrick, J. B., Aspidospermine II, *J. Amer. Chem. Soc.*, **76**, 5603 (1954).

Stork, G. and Dolfini, J. E., The total synthesis of *dl*-aspidospermine and of *dl*-quebrachamine, *J. Amer. Chem. Soc.*, **85**, 2872 (1963).

CHAPTER XII

PICRALIMA ALKALOIDS

MANY years ago a number of alkaloids (Chart 12.1), including akuammicine (Chart 6.1), had been isolated from *Picralima klaineana*. Until very recently they had resisted efforts towards the establishment of their nature. They, along with echitamine from *Alstonia scholaris*, because of their structures, have been responsible for bringing into sharper focus the still developing picture of indole alkaloid biogenesis. To crystallize what has been inferred in earlier chapters, a unified theory for the origin of these and related Type I alkaloids some as yet unknown is sketched in Chart 12.2. They can all be considered to be derived from a primary oxidation product, a radical, or a 7X-7H-indolenine where X is a good leaving group. Such indolenines have already been used to prepare oxindoles and pseudoindoxyls (Chart 5.7) and a 3,4-*seco* compound (p. 67) of representative tetrahydro-β-carboline alkaloids. It is almost certain that sooner or later equivalents of the hypothetical bases, e.g. alkaloid-ω will turn up in nature and it will be transformable into the picralima and strychnos skeleta.

The structures of the picralima bases (with the exception of echitamine determined by the X-ray method) depend on the sum total of the interpretation of the results of the physical measurements on the alkaloids and their transformation products as well as conversion into *nor*fluorocurarine and akuammicine.

From the point of view of the mass spectroscopist the mass spectra of a picralima alkaloid would be expected to mimic that of 2,16-dihydroakuammicine since the primary fissions would generate one and the same intermediate ion (Chart 12.3). This illustrates what is now a well recognized limitation of mass spectrometry in structure work but one which was not appreciated in the early stages of the method's application to indole alkaloid chemistry.

CHART 12.1
PICRALIMA TYPE ALKALOIDS

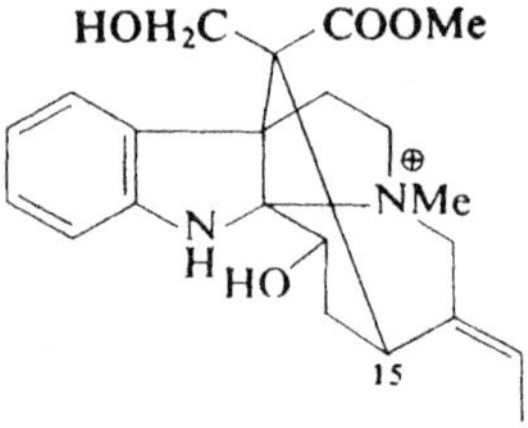

echitamine

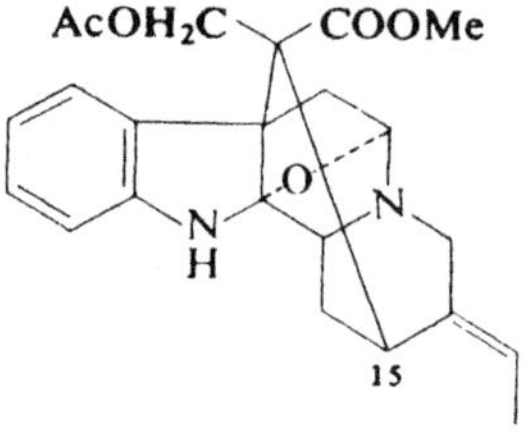

picraline

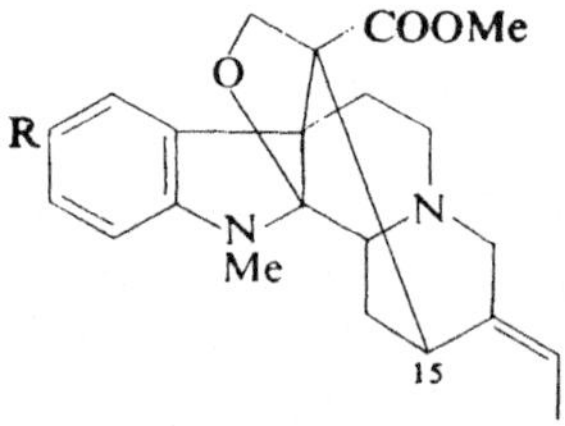

pseudoakuammigine (R = H)
akuammine (R = OMe)

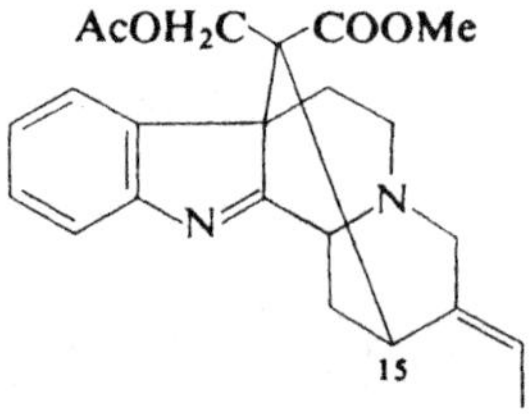

akuammiline

CHART 12.2
RELATIONSHIP OF THE PICRALIMA ALKALOIDS TO OTHER GROUPS

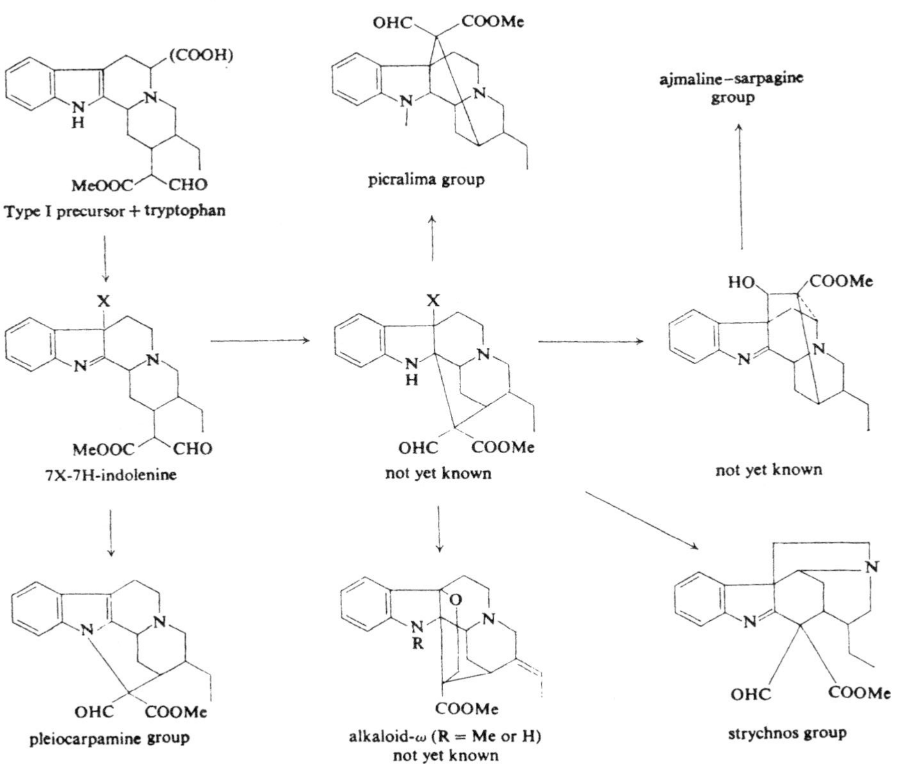

Chart 12.3
Fragmentation under electron impact of 2,16-dihydroakuammicine and pseudoakuammigol

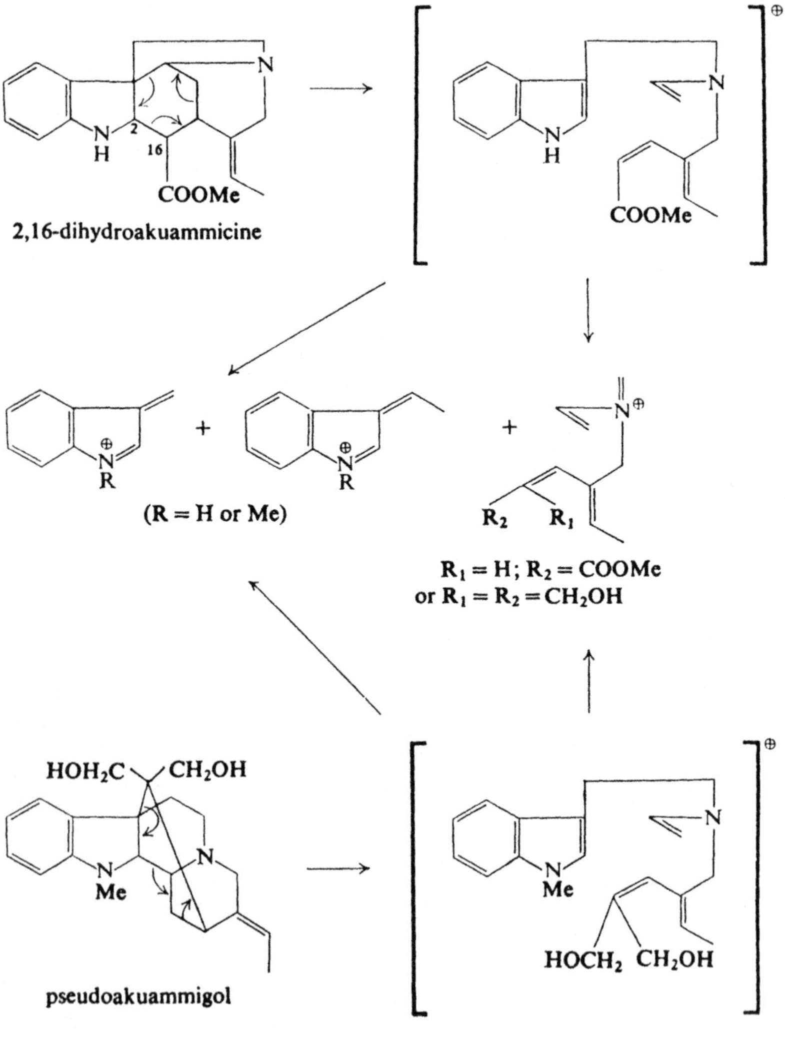

Picraline. When picraline was treated with acid it deacetylated and under more vigorous conditions with base a retrodealdolization took place with the additional loss of the hydroxymethylene group to furnish picrinine, lithium aluminium hydride reduction of which afforded the "alcohol". It was the attempt to carry out

CHART 12.4
CONVERSION OF PICRALIMA INTO STRYCHNOS BASES

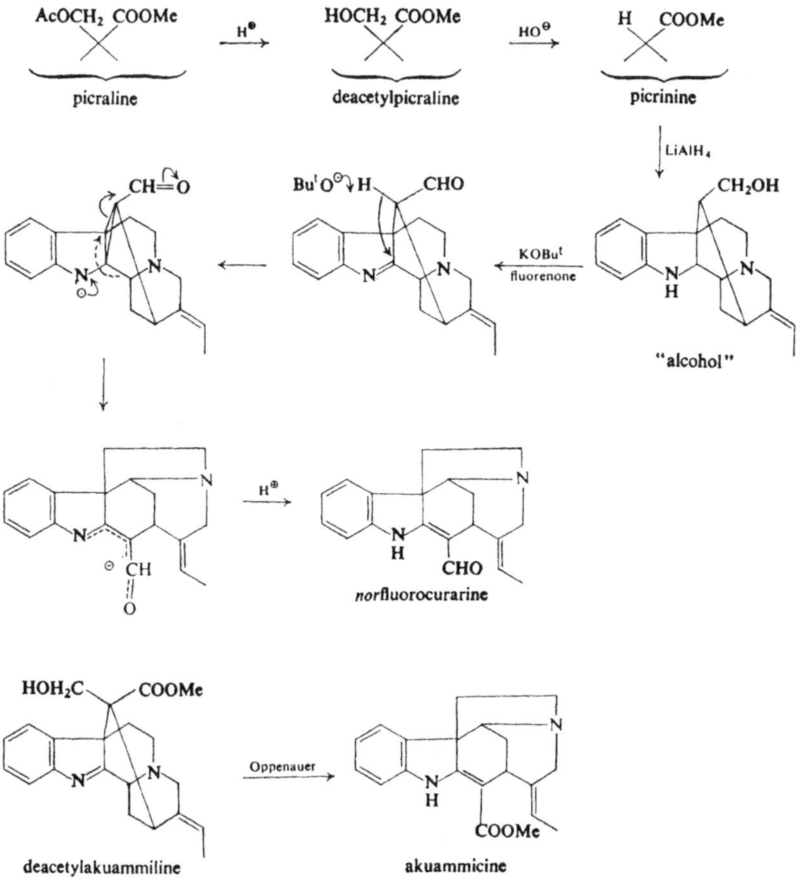

an Oppenauer oxidation on the latter compound which gave an indolenine aldehyde plus a surprising but exceedingly useful second product, *nor*fluorocurarine (Chart 12.4). By an analogous set of reactions deacetylakuammiline gave akuammicine. Both sets of transformations settled the absolute stereochemistry of the group which turned out to have the same configuration at C-15 as the rest of the Type I alkaloids.

CHART 12.5
ACID TRANSFORMATIONS OF PICRALIMA BASES

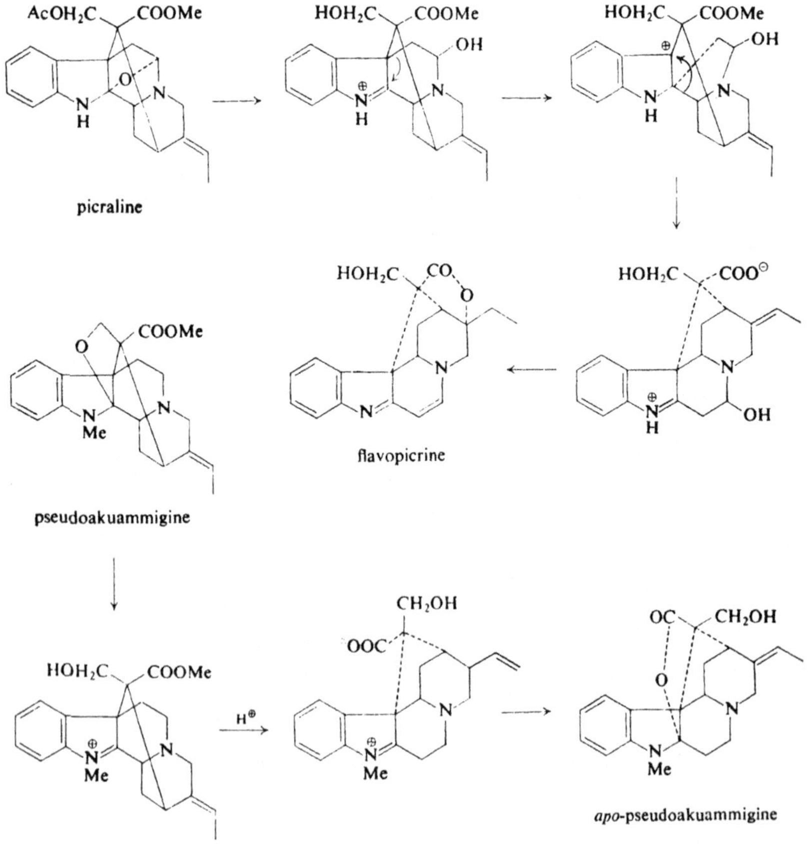

Picraline underwent a very interesting rearrangement upon warming in 3*N* hydrochloric acid to generate a yellow crystalline base, flavopicrine (Chart 12.5). The proposed structure for flavopicrine was in complete agreement with its analytical data, limited chemistry and physical measurements in particular the ultraviolet, infrared and nuclear magnetic resonance spectra. It should also be noted that this interpretation settled the orientation of the methoxycarbonyl group in picraline. Pseudoakuammigine under similar acidic conditions afforded *apo*-pseudoakuammigine where this time an easier lactonization occurred at C-2.

EPILOGUE

NATURAL products were the first available organic compounds and their variety and complexity has provided an ever present challenge to the chemist. However, the ability to isolate these materials does not necessarily mean that they exist as such in the living plant. Many of the substances obtained may be transformation products brought about by the simple act of drying or extracting the plant, or, in a more subtle way they may have been generated non enzymatically from labile precursors long after their deposition in a particular tissue. There are many examples in this book which provide circumstantial support for the latter thesis. The true answer to this question has an obvious bearing on the place of alkaloids in the plants' anabolic and catabolic metabolism and its outcome will help to put indole alkaloid biogenesis into proper perspective.

The application of the mass spectrometric method is leading to structural solutions with ever-decreasing amounts of material, so the need for rapid stereoselective syntheses of complex nuclei is becoming more urgent especially if we are to learn their chemistry and it is hoped that this summary of indole alkaloid chemistry will see its fulfilment in making some contribution to future work.

INDEX OF PLANT NAMES

INDEX OF CHEMICAL COMPOUNDS, ETC.

CHART 3.5

IMPORTANT REACTIONS OF INDOLES (SCHEMATIC)

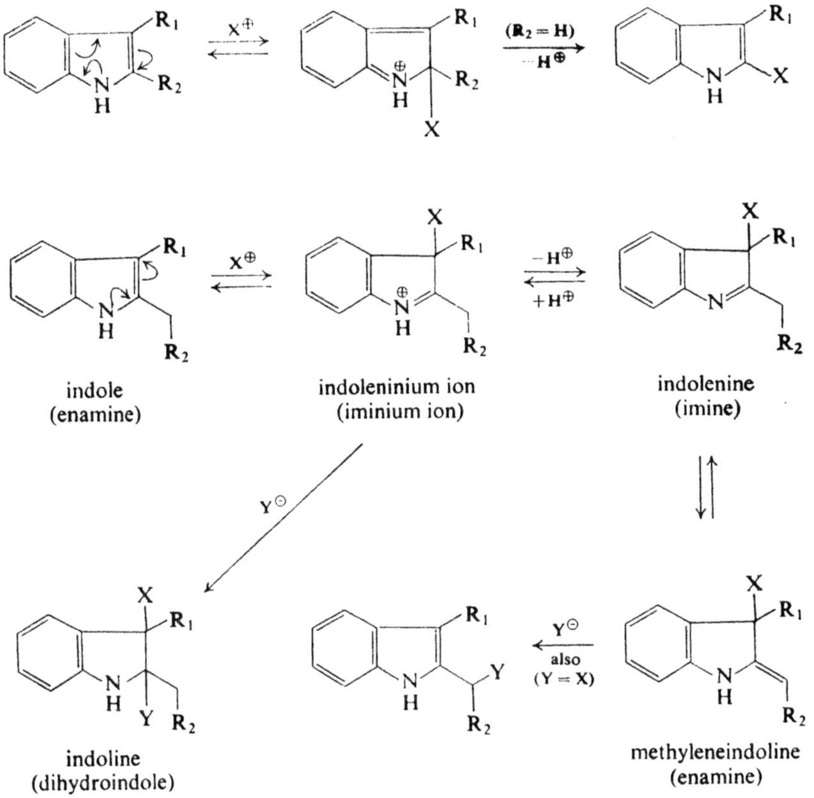

Printed in Dunstable, United Kingdom

91221480R00097